The Swan

THE SWAN

A Biography

Stephen Moss

◧ SQUARE PEG

1 3 5 7 9 10 8 6 4 2

Square Peg, an imprint of Vintage,
20 Vauxhall Bridge Road,
London SW1V 2SA

Square Peg is part of the Penguin Random House group of companies whose
addresses can be found at global.penguinrandomhouse.com

Text copyright © Stephen Moss 2021

Stephen Moss has asserted his right to be identified as the author of this Work
in accordance with the Copyright, Designs and Patents Act 1988

First published by Square Peg in 2021

Penguin.co.uk/vintage

A CIP catalogue record for this book is available from the British Library

ISBN 9781529110371

Typeset by Dinah Drazin

Printed and bound in Italy by L.E.G.O. S.p.A.

The authorised representative in the EEA is Penguin Random House Ireland,
Morrison Chambers, 32 Nassau Street, Dublin D02 YH68.

Penguin Random House is committed to a sustainable future for our
business, our readers and our planet. This book is made from
Forest Stewardship Council® certified paper.

*To my mother, Kay Moss, who first took me
down to the riverbank to feed the swans*

SWAN

Feathersung in blue,
you coast liquid holloways,
great wings nested.

Bone beaked and archaic,
you drift down skywaters,
cloud maps circled in your eye.

Sun winks and
galaxies explode about you.
Starfields of tiny lights
cloak you in constellation.

You are this river's dreaming,
sun coins strewn
at your breast.

Ruth Lawrence

Contents

PROLOGUE

The mute swan is Britain's largest bird and one of the world's heavier flying animals. It is as familiar to most people as the robin. It is graceful, confiding and much loved.

Janet Kear, *The Mute Swan* (1988)

When I first hear the sound, somewhere in the far distance, it doesn't seem to be made by a bird at all. A regular, rhythmic series of huffs and puffs, as if someone is overdoing their exercise regime, or fanning a fire with a set of old-fashioned bellows.

I look up into the wide, blue Somerset sky as two huge, white birds fly low over my head, before heading away towards the horizon. A pair of mute swans, contradicting their name. The sound I can hear is not, however, made vocally, but by their wings, as they beat them in a steady rhythm through the air.

This phenomenon is not easy to capture in words, as the ornithologist David Bannerman notes in *The Birds of the British Isles*:

> The noise made by the air passing through their great wings produces, as the *Handbook* puts it, 'a musical humming throb difficult to describe, but very distinctive and audible for a considerable distance'.

While on a visit to Abbotsbury Swannery in Dorset, described in his 1983 book *The Kingdom by the Sea*, the American author Paul Theroux chose a more graphic description:

I heard a sound – two sounds – a rapid sawing, a high muffled hee-haw, like the harsh hum of silk being woven in a clapping loom. It came closer, strengthened to a kind of breathing, though I could not place it. I listened and looked sharp, and I saw two huge swans flying low over the Fleet, beating their wings, tearing the air with them . . . When they were directly overhead, they sounded like two lovers in a hammock.

But whatever image the sound conjures up, it is certainly impressive: given the right atmospheric conditions, flying swans can be heard from more than a mile away. Just like more conventional bird songs and calls, this is a communication tool: helping the birds stay together when airborne, especially when it is dark or foggy. For us earthbound mortals, it draws attention to a creature we are never likely to ignore: the largest, heaviest, and arguably most familiar, British bird.

Along with the robin, and domestic creatures such as the goose and chicken, swans are one of those birds that seem to have been in our lives forever. From our earliest memories, when we were taken to the park to feed the ducks, these huge (and to a small child, frankly terrifying) creatures have embedded themselves deep inside our consciousness.

In lowland areas of Britain, where the vast majority of us live, swans are a more or less constant presence, as Bannerman notes: 'The mute swan in these islands is almost a part of the countryside, for wherever one may be in rural England a pair of them is almost sure to be in possession of the village pond or nearby lake.'

In towns, cities, villages and the wider rural landscape, wherever there is a stretch of water for them to find food and make their nest, you will come across swans. From village ponds to city park lakes, along narrow streams and broad rivers, and from grassy fields to coastal estuaries, they are almost everywhere.

Unlike other birds which, even when common and widespread, are often shy and elusive, swans are not just ubiquitous, but also very easy to see. That's partly down to their bright white plumage, but also because of their huge size: an adult male swan is almost twenty times as long, and over 200 times as heavy, as our smallest bird, the goldcrest.

Their large size also means that swans simply don't fear humans, at least in the way that most other birds do. This does, however, have its downsides: as we shall discover, swans are surprisingly vulnerable to deliberate or accidental injury and death. So while at first sight swans may appear to be invulnerable, their lives are often marred by these violent, and sometimes fatal, encounters.

The ubiquity of swans has created a host of myths and legends about them: some going back to ancient civilisations, others far more recent in origin.

Swans' habit of mating for life – or at least until one of the pair dies, after which the survivor will usually seek out a new partner – has led to them being closely linked with love, fidelity and purity (because of their whiteness). This, in turn, has inspired writers down the ages to create countless stories and works of art about them, from the Ancient Greek tale of Leda and the Swan to Tchaikovsky's ballet *Swan Lake*.

There are modern myths about swans, too: that they are completely silent (they aren't); that they all belong to the Queen (they don't); or that a single blow of a swan's wing can break a man's leg or arm (it really can't).

As I showed in my previous biographies, *The Robin*, *The Wren* and *The Swallow*, there is often a clear connection between the behaviour of the real, biological bird and its cultural counterpart: put simply, their habits frequently give rise to their mythology and folklore. And like robins, wrens and swallows, swans have a very special place in our affections.

Just as with the robin, whose attacks on rivals can be vicious and even deadly, the way we view the swan – as a symbol of love and purity – is often at odds with its day-to-day behaviour. Although they cannot actually break a human limb with a single sweep of their wing, they are known to be aggressive towards any other creature – including human beings – that dares to venture too close to them or their young. Being confronted by a male swan, hissing and opening its wings to look as big and scary as it can, is a pretty alarming experience. As we shall discover, the swan is a more complex, and less benevolent, creature than we might first assume. As the eminent ornithologist and swan expert Christopher Perrins rightly points out, 'not everybody likes swans'.

I grew up in the village of Shepperton: then in the county of Middlesex, but nowadays a leafy commuter suburb in north Surrey, on the western fringes of London. The River Thames was less than a mile from our home. My lifelong passion for birds goes back, according to my late mother, to when she took me down the road to feed the ducks (and presumably swans) beside the river at Laleham. Later on, my childhood playground was the local gravel pits, which were also home to a thriving popu-

lation of mute swans. Like most people, I simply took them for granted.

Today, I live on the edge of the northern half of the Somerset Levels, those vast, waterlogged flatlands bordered by the Mendip Hills to the north and the Polden Hills to the south. When we moved here from London, fifteen years ago, I had no idea I would be living in the mute swan's main British stronghold. Well over 1,000 birds – a significant proportion of the UK's population – spend the winter here, while each spring and summer, every watercourse seems to be home to a pair of swans, with their nest, eggs and fluffy young.

Whenever I take a walk or a cycle ride across the levels, whatever the time of year, I see swans. I may come across a group of non-breeding adolescents, loafing in a grassy field, or a breeding pair

– with or without their grey, fluffy cygnets – swimming along a rhyne (the local word for a drainage ditch). Even now, when no fewer than three species of egrets can be found here, any distant white bird is still most likely to be a mute swan.

Swans were not always so common in Somerset. Edwin Durston, a farmer featured in Robin and Romey Williams's book *The Somerset Levels,* talks of winters before the Second World War, when the moors were flooded and 'wild swans [i.e. Bewick's and whooper] came in', adding that 'there were none of they mute swans in them days'.

According to David Ballance, the acknowledged historian of the birds of Somerset, as late as the 1950s mute swans hardly ever featured in the annual *Report on Somerset Birds*. Yet records of the species being present in the county go all the way back to the middle of the thirteenth century, when King Henry III procured 'Somerset swans' for Christmas banquets, at the royal seat of Winchester Castle.

The apparent absence of mute swans from the record books might seem curious at first – until we discover that, surprisingly recently, mute swans were considered to be domesticated birds, hardly thought to be worth a mention in official accounts of wild bird sightings. No doubt the species has always been here in Somerset, and they are certainly abundant nowadays. During the time I have been writing this book, I have become even more aware of their presence.

While this biography mainly focuses on the mute swan – by far the most familiar member of its family – I shall also be writing

PLATE 156.—SWANS—MUTE-SWAN *(red)*; WHOOPER-SWAN *(middle)*; BEWICK'S SWAN: G. E. Collins

about some of the world's other swan species. Two of these – whooper and Bewick's swans – are autumn and winter visitors to Britain from their breeding grounds to the north and east. Known as 'wild swans', to distinguish them from their 'tame' counterpart, each has its own fascinating story to tell (see Chapter 5).

Another species of swan is also well known to a British audience. Native to Australia, the black swan is, as its name suggests, the mirror image of our familiar white swan. This opposition is embedded deep in our culture, as in the 2010 film *Black Swan*, which starred Natalie Portman as a ballet dancer struggling with her inner demons (see Chapter 1). The past decade or so has also seen the rise of a socioeconomic phenomenon known as the 'Black Swan Event' – meaning a hitherto unprecedented and apparently unforeseen catastrophe such as the 2008 stock market crash, or the Covid-19 pandemic.

Above all, though, I have chosen the mute swan as the main subject of the book because of its unique place in our hearts. Like the other subjects of my bird biographies, the swan means something special to millions of Britons and, like them, it is central to our day-to-day lives.

It would be fair to say that many people don't really think of the swan as a bird at all, but as a kind of 'national treasure': the avian equivalent of Sir David Attenborough, Dame Judi Dench or (rather aptly, given the species' royal connections) the Queen. But, as it turns out, there is far more to swans than this rather one-dimensional image would lead you to believe.

I

SWANS

What in nature can be more beautiful than the grassy-margined lake, hung round with the varied foliage of the grove, when contrasted with the pure resplendent whiteness of the majestic Swan, wafted along with erected plumes, by the gentle breeze, or floating, reflected on the glossy surface of the water, while he throws himself into numberless graceful attitudes, as if desirous of attracting the admiration of the spectator!

Thomas Bewick,
A History of British Birds (1804)

Swans are so common, so familiar, and so much a part of our daily lives, that only when we step back and take a closer look at them do we realise what extraordinary birds they really are. Of all the world's flying creatures, they are one of the largest and heaviest, and also very beautiful. Unlike the majority of wild birds, they are easy to see and often approachable. Get too close, though, and you will soon be sent packing, with a loud hiss and a lunge from that long neck and fearsome-looking bill.

Their tameness means that they have been captured in millions – possibly billions – of photographic images. Many of these confirm the verdict of author Sylvia Bruce Whitmore, that 'the proportions and posturings of the mute swan are more harmonious than those of the other swans', indeed, more so than most other birds. They appear to know it, too: as Jan Struther, author of the 1940 novel *Mrs Miniver*, points out, 'Swans always look as though they'd just been reading their own fan-mail.'

'Grace', 'beauty' and 'elegance' are the words most often used to describe mute swans. This comes from a combination of their shape and posture: the way they hold themselves, with that curved neck, the fluffed-out feathers and uptilted tail, so different in profile from the long, straight neck, flatter plum-age and lowered tail of their northern relatives. A well-known (though unattributed) epithet, dating back to the early nine-teenth century, aptly describes the swan as 'the peaceful mon-arch of the lake'.

It is the swan's apparent effortlessness, as it floats along a river or stream, that gave rise to the commonplace self-help concept that we should strive to be more like them: appearing calm and graceful on the surface, yet furiously paddling their legs beneath, to propel themselves along. In fact, swans do not need to paddle particularly strongly, even against a current.

On open water, they will take advantage of strong tailwinds to move at much higher speeds than they could achieve under their own steam: a technique known as 'windsurfing'. This was first discovered by a Swedish scientist who, on three separate oc-casions, at three different locations, observed mute swans using a following wind. They did so by arching their wings above their back to form sails, and then simply gliding along, without having to expend any extra energy.

In flight, too, mute swans are wonderfully impressive: wings held wide, neck outstretched, as they plough through the air in a dead straight line, while making that unique beating sound with their wings. Only when they come into land do they lose some of their poise and elegance. When landing on water, they beat their wings backwards, and hold their feet out in front of them, both of which help them to slow down for a safe landing.

Take-off is not easy either, for such a heavy bird. When taking flight from the water, a swan uses a combination of its wings and

legs to gain speed: beating the air with its wings, while simultaneously 'running' across the surface.

As the bird gains speed, the air passes over its aerofoil-shaped
wings (which, like an aircraft wing, are broader at the front than
at the back), producing low pressure above and high pressure
below. This in turn creates lift, which helps to combat the pull
of gravity that would otherwise prevent the bird from getting
airborne. Once the swan reaches the critical speed of around
21 kph (just over 13 mph), lift finally overcomes the opposing
force of gravity, and the bird can take to the air.

The similarity between swans and aircraft led, in the 1990s,
to a classic series of commercials for the Dutch airline KLM.
These used slow-motion footage of mute swans in flight, usually
accompanied by a music track, including the introduction to the
Thunderbirds TV series, Randy Crawford's song 'One Day I'll Fly
Away' and (rather oddly for a waterbird) Tom Jones's 'Green,
Green Grass of Home'. The best, however, was by far the sim-
plest: a recording of an air-traffic controller giving permission
for take-off, set to footage of the swan doing exactly that. The
analogy between swans and aircraft is a good one: both need a
runway, and both are cumbersome when taking off, yet graceful
and elegant once they finally get airborne.

Yet the gracefulness of swans, so obvious on the water and in
the air is, as Sylvia Bruce Whitmore notes, 'strangely lacking on
land'. When they walk along a riverbank or beside a lake, they
appear slow and awkward, their body swaying from side to side
as they do so. This is mainly because, as with other waterbirds,
the swan's legs are positioned well back on the body: ideal when

trying to gain traction while swimming, but far less useful when attempting to move on land.

The other main characteristic of swans is their huge size – they are one of the largest and heaviest birds on the planet. A fully grown mute swan really is spectacularly big. From the tip of its bill to the end of its tail, it measures over 1.5 metres – roughly 5 feet – and has a wingspan of 2.2 metres – over 7 feet. On average, males weigh 10.75 kg (nearly 24 lb), while the smaller females tip the scales at 8.5 kg (just under 19 lbs). However, the largest males can weigh as much as 15 kg (33 lb), while one Polish specimen is said to have reached an incredible 23 kg (over 50 lb). Such a bulky specimen would almost certainly have been unable to get airborne at all.

Even so, like all flying birds, swans are very light compared with mammals of a similar size and bulk. That's because their bones, along with the lower shafts of their feathers, are hollow and filled with air. A mammal of roughly the same length as a swan, such as a sheep or goat, tips the scales at between 50 and 100 kg – between five and ten times the average weight of an adult swan.

Mute swans are generally heavier than whoopers, but are still not quite the biggest member of their family. That honour goes to the trumpeter swan of North America, which is rivalled only by the Dalmatian pelican for the title of the largest waterbird on the planet.

Swans are members of the order Anseriformes, popularly known as 'wildfowl' or 'waterfowl'. The vast majority of these – about

170 species, including ducks and geese as well as swans – are in the family Anatidae.

Swans, geese and ducks share many characteristics. They are mostly gregarious birds, readily forming flocks, especially outside the breeding season; they have a strong attachment to their offspring, which remain with their parents long after leaving their nest; and they are mainly aquatic in their habits, though they do frequently feed or rest on land.

The Anseriformes is one of only two bird orders (the other being the Galliformes, or gamebirds) that were on Earth at the same time as the dinosaurs. Forebears of present-day swans – including the 99-million-year-old *Vegavis* – were present during

BEWICK'S SWAN.   MUTE SWAN.   WHOOPER SWAN.

the Mesozoic Era, and shared our planet with, among others, velociraptors, stegosaurus and the legendary Tyrannosaurus Rex. Somehow, the ancestors of today's ducks, geese and swans managed to survive the asteroid that devastated life on our planet roughly sixty-five million years ago and drove the dinosaurs to sudden and rapid extinction.

The genus *Cygnus*, which includes the six extant species of swan – mute, whooper, Bewick's (and its North American counterpart, the tundra swan), black, trumpeter and the black-necked swan of South America – evolved between five and twenty-three million years ago, during the geological period known as the Miocene, probably in Europe or western Asia. One other 'swan' species, the coscoroba swan of South America, is now thought to be more closely related to geese or shelducks.

Today, unlike the true geese, which are confined to the northern hemisphere, swans can be found in suitable habitats across much of the planet – they are only missing from Antarctica and sub-Saharan Africa.

The mute swan is found throughout the temperate regions of Europe, from southern Scandinavia in the north to Iberia in the south, and east to Russia, though it also occurs patchily across central Asia, and is a scarce winter visitor to North Africa. Mute swans have been introduced to North America – where they are often regarded as a pest – and there are small, localised populations in Western Australia, New Zealand and Japan.

The world population of the mute swan is estimated by BirdLife International at between 598,000 and 615,000 individuals, which, given that mute swans do not breed until they are between three and seven years old, equates to between 83,400 and 116,000 breeding pairs. The good news is that numbers appear to be on the increase, and the species is not considered to be globally threatened.

Mute swans are doing pretty well closer to home, too, being found across much of lowland Britain, in many town and city centres as well as the wider countryside. The species is either scarce in or absent from most of the uplands (above 300 metres, roughly 1,000 feet), including the Pennines, Dartmoor and Exmoor, central Wales and much of northern Scotland.

Mute swans can also be found on all our major offshore island groups, including Scilly, the Western Isles, Orkney and most recently Shetland, having colonised Britain's northernmost archipelago in the early 1990s. As a result, they now overlap as a breeding bird with the whooper swan, which expanded its range south from Iceland to Shetland at about the same time.

Not surprisingly, the mute swan's strongholds coincide with our largest and most extensive wetlands: the broads and fens of East Anglia, the lochans of the Outer Hebrides, Orkney, major river systems such as the Thames, Trent and Severn and, as I have already noted, my home patch of the Somerset Levels.

Numbers fell to just 3,500–4,000 breeding pairs in the mid-1950s, and perhaps even fewer in the 1960s, following the consecutive freeze-ups of 1961/62 and 1962/63, which led to the deaths of roughly one-quarter of all Britain's mute swans. In a double whammy, poisoning by lead fishing weights was also causing heavy mortality. During the following decades, numbers slowly began to rise again, partly thanks to a ban on lead weights from

1987, and also because winters have become on average milder. An increase in suitable wetland habitats, including the creation of gravel pits, is also likely to have helped the species.

Long-term surveys carried out by the British Trust for Ornithology (BTO) show a steady increase in both breeding and wintering birds, from the mid-1980s to around the turn of the millennium, since when numbers have been more or less stable. Today, there are between 6,300 and 7,600 pairs of mute swans breeding in the UK, more or less twice the number found here during the years after the Second World War, and as many as 53,000 individuals present in winter. Good news for all of us who admire these beautiful birds.

It could be argued that the mute swan is so distinctive, and so well known, that it hardly needs describing at all.

The opening lines of various field guides bear this out. 'Huge and white . . .', 'The familiar swan . . .', '*Huge*. Plumage *white*. Neck very long, head small . . .' and so on; while in their excellent 1986 monograph on the species, Mike Birkhead and Christopher Perrins apologise to their readers for needing to describe such a familiar bird at all.

Yet given that there are two other species of swan which can be confused with our more familiar species, it is worth going into a little more detail about what makes the mute swan unique.

The plumage is indeed white – at least in adults – but in nature, as in paint charts, 'white' comes in many different shades. When I scan the wet fields and marshes near my home, I can usually pick out distant swans from the now frequent great white and little egrets because, although the swans are white, they are not quite as 'snow-white' as those slim, elegant birds.

In reality, the mute swan's plumage is a kind of off-white or, using the language of light bulbs, 'warm white', compared to the brighter, colder shade of the egrets. Swans also often show a distinctive rust-coloured shading around their head and neck. This comes from staining by iron and tannins, which are present in the water into which the birds dip their head and neck to feed.

White might seem an unusual colour for a bird – after all, it does make swans very conspicuous. But as David Cabot points out in his 2009 Collins New Naturalist volume *Wildfowl*:

> The mute swan . . . is so large and powerful that it has no need to conceal itself. In fact, the white plumage confers an advantage in advertising the birds' presence during the breeding season, thus warning off other swans from their territory.

The other distinctive feature – and the one that conclusively separates an adult mute swan from the other two British species – is that its bill is a deep orange colour, with black on the tip, rather than the combination of yellow and black shown by both Bewick's and whooper swans. Mute swans also have a noticeable black 'knob' at the top of their bill, which gives the Dutch their name for the species: *knobbelzwaan*. The size of the knob is the best way to distinguish the sexes: the male, known as the 'cob', has a larger and more protruding one than the female (the 'pen'), giving him a rather sterner appearance. Both male and female have dark greyish legs and feet.

1. BEWICK'S SWAN. C. bewicki.
2. ALPHERAKY'S SWAN. C. minor.
3. THE WHOOPER. C. cygnus.
4. THE MUTE SWAN. C. olor.
½ not size.

Once they fledge from their fluffy down into their first juvenile plumage, about four to five months after they hatch, young swans (known as cygnets) are mainly greyish-brown. By the time they are a year old they have usually turned mainly white; however, they retain their greyish bill until they finally reach maturity at the end of their second summer or autumn, up to eighteen months after they were hatched.

'Swan' is, along with 'goose' (but strangely not 'duck'), one of the oldest of our common bird names. It has not changed its spelling since it was first recorded in English during the Anglo-Saxon era, and the same word (with variant spellings) is used in German (*Schwan*), Dutch (*zwaan*), and the various Scandinavian languages (*svan* or *svane*). The similarity between these modern names reveals that the original must go back to a time when the split between the Germanic and Scandinavian languages had not yet occurred: i.e. before 500 BC. In fact, it probably arose far earlier than that.

It has been suggested that the word 'swan' originally comes from a Sanskrit word meaning 'sound', which could of course refer to the distinctive calls made by the whooper swan, but might instead derive from that extraordinary throbbing sound of the mute swan's wings.

And what of the epithet 'mute'? First applied by Thomas Pennant in 1785, this familiar adjective is, in many ways, misleading. True, the species does not call in flight, as do Bewick's, whooper and trumpeter swans (the last two being named after their flight call). But as anyone who has got close to mute swans can testify, they have a wide vocabulary of vocal sounds, ranging from a muffled bugling call when defending their territory, to what has been described as 'an explosive snorting or hissing' when under threat from humans or animal predators. Male and female mute swans make a range of other sounds, too: males snort, while females gather their young using a sound like a yapping puppy. Cygnets may also make a high-pitched peeping call if they become temporarily separated from their parents.

*

The story of the mute swan in Britain is a long, eventful and often surprising one; indeed, its history is very different from that of any other British bird. There is – as with so many aspects of swans' lives – a potent mixture of truth and error, myth and legend, speculation and rumour, and what we might nowadays call 'fake news'.

Let's start with one of the easiest myths to debunk: the date of their first arrival on these islands, which is often given as the last decade of the twelfth century. In 1192, Richard the Lionheart (who, given that he never actually learned to speak English, should actually be known as *Richard Coeur de Lyon*) was returning from the Third Crusade to the Holy Land. On his journey home, he is said to have picked up a few pairs of mute swans – supposedly from the island of Cyprus – and brought them back to Britain.

It's a plausible tale, but unverifiable, and almost certainly untrue. It does not even surface in print until some 650 years later, in William Yarrell's *A History of British Birds*, published in 1843. The truth is that, on his way home from the crusades, Richard was captured and taken prisoner by his enemy Leopold, Duke of Austria, and only released when a ransom of 100,000 marks – a vast sum, equivalent to roughly twice the annual income of the English Crown – was eventually paid.

As N. F. Ticehurst notes in his 1957 study *The Mute Swan in England*, it is highly unlikely that, under such difficult and testing circumstances, Richard would have been thinking about swans:

He could of course have sent back a few couples of swans on his returning supply ships, but it would have been a most hazardous proceeding and it seems doubtful if they could have survived such a long, difficult and dangerous voyage in the small ships then in use.

And, as Ticehurst pointedly adds, 'What could have been the object in importing them from Cyprus, when they could have been obtained with far greater ease and at far less expense from the opposite side of the North Sea?' Most telling, though, is the evidence that mute swans were kept by the Bishop of Lincoln at his park at Stowe, a few miles north of the city, in 1186, six years before Richard returned from the Third Crusade.

More circumstantial evidence that the story is untrue can be found in records from the reigns of Richard's successors as kings of England, specifically relating to the procurement of swans for the extravagant feasts that were such an important part of royal life. As I have already noted, in 1247 Henry III ordered forty Somerset swans to be procured for his Christmas feast at Winchester; two years later, that figure had risen to more than a hundred. During the following years, orders were issued to abbots, priors and bishops in many parts of England, to obtain as many swans as they could, to provide for the royal festivities. As Ticehurst points out, the idea that, half a century or so after Richard I was supposed to have first brought mute swans to England, the species could have expanded its range and numbers rapidly enough to supply so many birds for these annual royal feasts, renders the crusade theory 'entirely fallacious'.

Place names provide more evidence that swans have long been native to Britain. Swanage in Dorset, Swanscombe in Kent, and of course Swansea in south Wales would all seem to have been named after this familiar bird, but beware: it is likely that many, if not most, of these names are corruptions of earlier, Anglo-Saxon words, which would have had very different meanings. 'Swansea', for example, is thought to derive from 'Sveinsey' – meaning 'Svein's island'; while other places with 'swan' in their name may come from an Anglo-Saxon word meaning 'swineherd', or 'swain', meaning a callow young man, rather than the bird. Nevertheless, the Swedish place-name scholar Eilert Ekwall suggests that a name such as Swanbourne (in Buckinghamshire) – meaning 'swan stream' – does indeed derive from the bird.

Moreover, as James Fisher points out in his masterly survey of Britain's birdlife, *The Shell Bird Book*, the bones of mute swans dating back to Neolithic times (between 10,000 and 3,500 BC) have been found in peat bogs across much of southern Britain, including a site excavated by archaeologists in Cambridgeshire. Here, on what would have been a vast, watery fenland, they lived alongside an astonishing range and number of species including the Dalmatian pelican – one of the very few flying birds to surpass the mute swan in size.

Mute swans were found in Roman Britain too: there is evidence of their presence in archaeological deposits from the Romano-British city of Camulodunum (present-day Colchester) in Essex. And they were also known to a later cohort of invaders, the Anglo-Saxons, long after the Romans had left the country.

The *Exeter Book*, dating from the end of the first millennium, contains virtually all that survives of the writings and poetry of this enigmatic era. In its pages, an anonymous and now long-forgotten poet celebrates not just 'swans' (which could of course refer to one of the two species of 'wild swan') but specifically the mute swan. He does so in one of the many riddles contained in the book, which were written to test and amuse an educated and knowledgeable readership.

The original Anglo-Saxon is now incomprehensible to all but a few scholars; fortunately, one of them, Dr Richard Fahey of the University of Notre Dame, in Massachusetts, has translated the riddle into modern English:

> My dress is mute, when I tread the ground, or occupy my abode or stir the water. Sometimes my garments and this high wind heave me over the city of heroes, and then the strength of the skies bears me far and wide over the people. My adornments then resound loudly, and ring, and sing brightly, when I am not hanging on sea or land, a travelling spirit.

As a description of the sound made by a mute swan's wings in flight, this is hard to beat. It also clearly rules out the possibility that these swans, recorded several centuries before Richard I's reign, were of the whooper or Bewick's varieties, neither of which makes the same distinctive sound with its wings when passing overhead.

Dr Fahey is not the only scholar to have translated the *Exeter*

Book riddle. The poet and naturalist Geoffrey Grigson did too, keeping more of the rhythm, metre and alliteration of the original verse:

> Clothes make no sound when I tread ground
> Or swell in dwellings or disturb the flow.
> Gear and lofty air at times
> Above men's towns will lift me:
> The brisk breezes bear me far, and then
> My frettings loudly rush and ring
> Above the folk and clearly sing
> When I forth-fare on air and know
> And feel, no fold or flow.

Grigson, a contemporary of James Fisher, could not resist lobbing a gentle jibe at his friend's more scientific interest in birds:

The answer to that riddle was a swan, of course: the white bird, the bird of the long neck, the bird that floats and sees its own image in the water, bird of gods and poets and kings. Of myth and legend, madrigal and opera. Not, my dear ornithologist, with your prosy ways and your prosy field-glasses and your prosy statistics, not your kind of bird.

Although said in jest, to tease his old friend and rival, Grigson's comment contains more than a grain of truth. Even today, birders still tend to regard mute swans as 'different' from other birds: somehow not quite so wild, and therefore not quite as genuine.

This is not an entirely modern attitude: it dates all the way back to the medieval period, when swans were at least partly domesticated and, more specifically, linked to the all-powerful monarchy. This set them apart – and does still – from any other British bird.

David Cabot takes a rather jaundiced view of the way we treat the species today:

British and Irish mute swans have assumed an almost pet-like status, many existing on a diet of bread and other scraps proffered at ornamental lakes in public parks, and on ponds, rivers, canals, harbours and estuaries. Our mute swans are among the most mollycoddled of birds, and have adroitly exploited their symbiotic

Cygnus olor (Gmelin)
DE KNOBBELZWAAN ♀

relationship. In exchange for plentiful supplies of food . . . and generally predator-free nesting sites, they grace our public waterways with their elegance and majesty to provide endless hours of aesthetic enjoyment.

But how did this 'symbiotic relationship', so different from that we have with any other British bird, come about? To find out, we must delve back into the long history of the unique, and very special, relationship between human beings and swans.

The Scottish ornithologist William MacGillivray, a contemporary of William Yarrell, was scathing about the notion that the mute swan could ever be regarded as anything other than a domesticated creature. As a result (and with a rather snide reference to his rival), he did not include it in his own five-volume work, published in 1852:

The Common Tame Swan . . . is admitted by Mr Yarrell into the series of British Birds; but, as there appears no evidence of its ever having been shot or caught, in a truly wild state, in any part of Britain, I am constrained to omit it.

MacGillivray was far from alone in his views: the species was widely known during this period as the 'Tame Swan', and was referred to by that name as recently as 1895 by the noted ornithologist W. H. Hudson. Like others, Hudson regarded the swan in much the same way as we might view the chicken, turkey or farmyard goose today: a domestic bird kept solely for the purpose of producing meat for food and, in the case of the swan, also feathers, which were large and sturdy enough to be made into quill pens.

However, writing a century earlier, Thomas Bewick had taken issue with the notion that the swan was somehow equivalent to other farmed animals:

> This species cannot properly be called domesticated; they are only as it were partly reclaimed from a state of nature, and invited by the friendly and protecting hand of man to decorate and embellish the artificial lakes and pools which beautify his pleasure grounds. On these the Swan cannot be accounted as captive, for he enjoys all the sweets of liberty.

Bewick also noted that swans do not take readily to captivity:

> The Swan will not thrive if kept out of the water: confined in a court yard, he makes an awkward figure, and soon becomes dirty, tawdry, dull, and spiritless.

The mute swan has long been known as a 'royal bird', with the Crown having the right to own every unmarked swan in England. This was formalised by Edward IV in 1482, in a royal statute, the Act for Swans. However, by then it had already been the custom for at least 300 years: the phrase appears in a work by the late-twelfth-century scholar Giraldus Cambrensis (also known as Gerald of Wales).

The first written evidence for swans being considered 'royal birds' comes much earlier: from a full century before the Norman Conquest. In the year 966 AD, the abbots of Croyland (now known as Crowland), a settlement in the south Lincolnshire fens, were granted the right by King Edgar to keep 'stray swans', which could possibly suggest that other swans were not 'stray', but had legal owners.

The royal connection has led to some confusion, which continues today, with many people still believing that the Queen owns all Britain's swans. Technically this is true for the whole of England, but in practice the rule only applies to any unmarked swans on the River Thames, which are caught and checked each year in a custom known as 'swan-upping' (see Chapter 3). Elsewhere, in Wales, Scotland and Northern Ireland, swans are not, and have never been, a royal bird.

For much of English history, though, the Crown was able to decide who and who could not keep swans – a prerogative it exercised throughout the late medieval period, from the twelfth through the sixteenth centuries, and some time beyond. The right to own swans was given to lords of the manor, the heads of religious houses, and other local dignitaries. As Professor Katy Barnett of Melbourne Law School points out, 'It might seem extraordinary for monks to have rights to swans, but the description of the monk in the General Prologue to Chaucer's *Canterbury Tales* suggests it was not unusual': 'A fat swan loved he best of any roost [roast].'

The 1482 Act banned the keeping of swans by 'Yeomen and Husbandmen, and other persons of little Reputation', the medieval equivalent of today's middle and working classes. The only people permitted to keep the birds were the nobility and landed gentry, who needed to own 'Lands and Tenements of Estate of Freehold to the yearly Value of Five Marks above all yearly Charges'. A mark was a notional unit of currency equivalent to 160 pence (two-thirds of a pound), so five marks would have been worth three pounds, six shillings and eight pence (roughly £2,200 at today's values).

The high economic value of swans meant that crimes against them – such as killing the birds or taking their eggs – were taken very seriously indeed: so much so, that they were classed as treason. During the reign of Henry VII, in the late fifteenth century, taking a single swan's egg was penalised with a year's imprisonment. Even today, killing a mute swan can be punished by a fine of up to £5,000 – far more than for causing the death of any other wild bird.

Norman Ticehurst, whose book is a treasury of intricate and fascinating detail, points out that if swans were already well established here as a wild bird before the onset of royal ownership, they would first have had to be captured before they could be farmed on a large scale. This would have been done, he speculates, by surveying an area each spring and summer for swans' nests – which are hardly inconspicuous – and then taking the cygnets before they fledged. They would then be hand-reared and pinioned: surgically removing the pinion joint on the bird's wings so they were unable to fly away.

Over the course of time it would easily have been possible to transform what had been a wild, free-flying population into a more controllable one. But although mute swans were tamed, they were never fully domesticated, as the ornithologist Dr Janet Kear points out in her 1990 book *Man and Wildfowl*:

> Domestication implies that man decides on the choice of mate, and thereby selects the genetic characteristics of the animal's descendants. For the swan, pinioning must have restricted the natural choice somewhat; however, man never seems to have stopped particular pairs breeding . . . and seldom selected between the cygnets that were to be taken fattened to the table. Youngsters that grew on to become the parents of the next generation were left to cope with all the hazards of the natural environment and its selective pressures.

This unusual, semi-feral status may have proved to be the salvation of the species. As Ticehurst notes, many other once-common species of waterbird – including the crane and bittern – were hunted to the brink of oblivion (in the crane's case, out of existence in Britain) during this period. The swan, however, was not. As Professor Christopher Perrins – an eminent ornithologist and, since 1993, the official Swan Warden – points out, 'they survived because people claimed ownership of them, so they had a vested interest in ensuring they left enough to maintain the breeding stock'.

Swans had some obvious advantages over keeping other waterbirds: having been pinioned, and so unable to escape, they could be released onto a suitable waterbody and simply left to their own devices. Unlike, for example, domestic chickens, they did not have to be fed all year round, as for much of the year they were able to find plenty of food for themselves. The adults were often given grain to supplement their diet in the winter, while the cygnets – destined for the table – were put in separate pens and fed on oats, barley or malt. However, unlike other waterfowl, such as ducks and geese, the adult birds could not be kept in flocks, as they are too territorial.

In 1698, the pioneering travel writer Celia Fiennes made a journey on horseback around the country, which she later published as *Through England On a Side Saddle*. While riding across the fens, near the cathedral city of Ely, Fiennes reports seeing breeding swans: 'Here I see the many swans' nests on little hillocks of earth in the wett ground that they look as if swimming with their nest.'

Owners of swans made handsome profits. In 1553, Sir William Cecil (the chief adviser and counsel to Queen Elizabeth I) sold thirty-nine cygnets for almost five pounds, which, after subtracting costs for food, and six shillings and eight pence paid to his swanherd, produced a profit of four pounds and two shillings – equivalent to more than £1,100 in today's money. Swans were also far more expensive than other birds: in the late thirteenth century they were priced at three shillings each (over £100 today), compared to a few pence for other species – so were only kept and eaten by the very rich.

Swans also became a status symbol: having these 'royal birds' on your land would, in Ticehurst's words, 'lend a certain air of distinction, well-being and importance to their owner'. Or, as

DOMESTIC SWAN.
Cygnus mansuetus (Gml.)

the writer Emily Cleaver puts it: 'Swans were luxury goods in Europe from at least the twelfth century onward; the medieval equivalent of flashing a Rolex or driving a Lamborghini.'

They were often given as gifts from one important person or institution to another, as in June 1552, when the City of Lincoln presented the visiting Duke of Northumberland with, among other items including cranes, bitterns, godwits, knots, pikes, breams and tenches, 'six Signetts'. As Christopher Perrins notes: 'Swans had a "snob value" far above simply that of their meat – it was a feather in your cap to have been granted the right to own swans, as it had been given by the monarch.'

Swans were also sometimes used to raise royal revenues. The National Trust property of Lytes Cary Manor in Somerset was built in the 1340s on land originally granted to William de Lyte by King Edward I in 1286. In return, the King demanded rent of either ten shillings a year (roughly £320 at today's values), or the contents of a swan's nest (presumably in the form of cygnets, rather than eggs) from the River Cary, which runs alongside the land. The latter option was presumably chosen, as the family's coat of arms, as shown in the manor's stained-glass windows, is three swans and a chevron, the latter signifying the river itself.

Elsewhere in Somerset, at the church of St John the Evangelist in the village of Countisbury, near Minehead, there is a medieval bench-end with a notable carving of a swan chained to a crown. Thought to date from the late fifteenth century, this heraldic emblem also appears on the county flag of Buckinghamshire, apparently – according to the cultural critic Professor Axel Goodbody – because swans were bred there for the king.

Professor Terry Gifford wrote a poem about it, 'The Chained Swan'. The last stanza reads:

> Beyond the isolated church of Countisbury
> the coastal path drops from Butter Hill,
> high above the sea, balancing through bracken,
> airy, loose and steep, towards the Lyn's mouth.
> Raven calls black-flap below. But from that church
> echo the distressed cries from a medieval bench-end
> of a swan chained to the crown slipped down her neck.

*

In the King James Version of the Old Testament, the swan is listed as an 'unclean' animal, whose consumption was – and for many practising Jews still is – prohibited under the laws of Judaism. However, the original reference to the swan from the Book of Leviticus, part of a long list of prohibited birds including eagles, vultures, kites and ravens (many of which are habitual scavengers, and so would have been dangerous to eat), is thought to be a mistranslation from the original Hebrew of some other, unknown bird species, possibly an ibis.

Swans (or specifically, the more tender cygnets) have long been prized for their tasty flesh, which has been described by a contemporary American chef, Mario Batali, as 'deep red, lean, lightly gamey, moist, and succulent'. Not everyone agreed: one eighteenth-century observer describes the meat as 'blacker, harder and tougher [than goose], having grosser Fibres hard of Digestion, of a bad melancholic Juice'. Thomas Bewick notes that they are 'by most people accounted a coarse kind of food', though he goes on to say that 'Cygnets are still fattened for the table, and are sold very high, commonly for a guinea [21 shillings, or about £45 at today's values], and sometimes for more.' Charles Darwin is also rather ambivalent, describing the swan's meat as 'pretty good, but tasted like neither flesh nor fowl, but something halfway like venison with wild duck'.

By the time Darwin was tucking in to his roast meal, swans were no longer as popular as they had been: the rise in the wages of farm labourers meant that keeping swans was costlier than before, while their inconvenient habit of being territorial limited the number that could be kept in one place. Other birds, such as the goose and the turkey (newly imported from the American colonies), were cheaper and more convenient to raise.

Being so large, a swan always requires extensive preparation before cooking, as this recipe for baked swan, from the 1670 cookbook *The Queen-like Closet or Rich Cabinet*, by the popular cookery writer Hannah Woolley, explains:

> To bake a Swan Scald it and take out the bones, and parboil it, then season it very well with Pepper, Salt and Ginger, then lard it, and put it in a deep Coffin of Rye Paste with store of Butter, close it and bake it very well, and when it is baked, fill up the Vent-hole with melted Butter, and so keep it; serve it in as you do the Beef-Pie.

After cooking, swans were often 'dressed' in their plucked feathers to create a more pleasing appearance when served. This was graphically described in a fourteenth-century French cookbook, *Le Viandier de Taillevent:*

> Take the swan, inflate it between the shoulders, slit it along the belly, and remove the skin (including the neck cut close to the shoulders). Leave the feet attached to the body. Put it on the spit, bard it, and glaze it. When it is cooked, reclothe it in its skin, with the neck very upright on the plate.

At any medieval feast, the pièce de résistance would be the equivalent of the modern 'turducken' (a chicken stuffed inside a duck, in turn stuffed inside a turkey) – but with no fewer than *eight* smaller birds inserted into the swan.

For a sheer frenzy of decadence, gluttony and conspicuous consumption, however, nothing can compare with the gargantuan blow-out held by one of the most powerful politicians of his or any other era, George Neville, in September 1465. Held to celebrate Neville's enthronement as Archbishop of York, the feast took place to the south of the city at Cawood Castle, and was the medieval equivalent of an all-you-can-eat buffet, lasting for several days. In his account of this epic event, from *The Brothers York: An English Tragedy*, the historian Thomas Penn vividly depicts the preparations involved:

> There was a frenzy of butchery. The carcasses disembowelled, plucked, prepared and dressed for the occasion included 1,000 sheep, 500 deer of various varieties, 400 swans, 2,000 each of pigs and chickens, 104 peacocks, 12 'porpoises and seals' and 6 wild bulls. An army of pastry-makers produced nearly 10,000 pastries, pies and tarts, as well as 'sugared delicates and wafers' with which the more privileged of the sated guests would idly toy as they digested the 22 pounds of meat that, on average, each was served. All this was washed down with 300 tuns of ale and a hundred tuns of wine.

'A tun', Penn helpfully informs us, 'is the equivalent of 256 gallons', which means that each of the 2,500 guests was supplied with almost 250 pints of beer and 66 bottles of wine, most of which they presumably drank. Along with the swans, they also consumed 200 cranes and 200 bitterns – more, in each case, than currently occur in the wild in the whole of Britain.

Incidentally, George Neville did not, as we might imagine, die from this grotesque surfeit of wine and waterbirds; instead, having conspired against his monarch Edward IV in favour of the temporary occupant of the throne Henry VI, he was arrested for treason and sent into exile in France. Eventually Edward relented, and Neville was permitted to return home to Cawood Castle, where he died of natural causes in June 1476.

Although the consumption of swans never again matched the excess of these medieval feasts, recipes for roast or baked swan continued to appear well into the Victorian era. However, by the time Mrs Isabella Beeton published her celebrated *Book of Household Management* in 1861, she explains that the bird was no longer regularly cooked and eaten. This may have been because the gamey taste of swan was beginning to fall out of favour, or because cygnets were no longer being raised for the pot.

But, according to Dr Steve Loughnan of the University of Edinburgh, there may be more complex reasons behind our reluctance to eat swans. In his 2014 paper 'The Psychology of Eating Animals', Dr Loughnan discusses the 'meat paradox' – why we choose to eat some animals, while others are considered taboo. He explains that:

> Biologically, there is obviously not much difference between a swan and, say, a duck or a goose. The line which says one is edible and one is not is thus arbitrary – ducks and geese belong to the 'food' category' and swans belong to the 'wildlife' category.

Roughly at the time swans fell out of fashion as a food item, the term 'tame swan' had been replaced by the name by which we know the species today, mute swan. Is it purely a coincidence that at the moment we started to consider swans as a truly wild species, we stopped eating them?

Despite this, *Mrs Beeton's All About Cookery*, published posthumously in Australia in 1923, more than half a century after the death of its eponymous creator, does include a recipe for roasted or baked *black* swan, stuffed with beef mince, layered with bacon rashers and covered in lemon juice, which is deemed 'sufficient for eight persons'.

Today, recipes aside, the black swan has found fame far beyond its native Australia. This is largely thanks to two unrelated but equally fascinating cultural phenomena: a socio-economic theory and an Oscar-winning movie. But before I examine these, let's take a closer look at the bird itself.

As its name suggests, the black swan is the mirror image of the familiar white swan we know so well. At a distance it does appear black, but a closer view reveals that the feathers on the wings and body have pale fringes, giving the bird a less uniformly dark appearance. The legs and feet are grey, and the bill is mostly blood-red, with a pale tip. When the black swan takes flight, it reveals snow-white feathers on the wing, a striking contrast to the dark shade of the rest of the plumage.

Black swans are native to Australia, where they are found across much of the country. They are most common in the south and east, and also in the southern part of Western Australia, where the species is the state bird, appearing prominently on the official coat of arms.

Once considered to be largely sedentary, black swans are, like many other Australian birds, now known to be nomadic, responding to seasonal rainfall by moving into new areas – including the usually arid interior of the country – to breed. With roughly half a million individuals – about the same global population as the mute swan – the species is not considered to be threatened.

Over the past century or so, black swans have been successfully introduced to New Zealand, while feral birds can also be found in Japan, China and Britain where, during the Victorian era, they were commonly kept on lakes in city parks. This sometimes led to conflicts between them and the local mute swans, as in this account from William Yarrell of a grisly incident in London's Regent's Park:

The two White Swans pursued the Black one with the greatest ferocity, and one of them succeeded in grasping the black one's neck between its mandibles, and shook it violently. The Black Swan with difficulty extricated itself from this murderous grasp, hurried on shore, tottered from the water's edge a few paces, and fell, to die . . . with great agony . . . Its foes never left the water in pursuit, but continued sailing with every feather on end, up and down towards the spot where their victim fell, and seemingly proud of their conquest.

Throughout the following century, black swans continued to be popular. Winston Churchill even kept a flock at his home at Chartwell in Kent, having been given them by Sir Phillip Sassoon (cousin of the poet Siegfried), a prominent politician and socialite.

Churchill became devoted to the birds, and claimed to be able to converse with them in their own language, which he dubbed 'swan-talk'. However, one of his bodyguards, Ronald Golding, discovered that the swans would actually respond to anyone who approached them:

It was some time after this discovery that I was walking down to the lake with Mr Churchill. I was a little in front, and watched carefully for the critical spot. I then called out in 'swan-talk' and the birds dutifully replied. Mr Churchill stopped dead. I turned round and he looked me full in the eye for a moment or two. Then the faintest suspicion of a smile appeared and he walked on in silence. No comment was ever made that this secret was shared.

In his autobiography, *The Eye of the Wind*, the conservationist Peter Scott reveals that the male of the pair at Slimbridge came from Churchill's collection:

> We heard that there was some overcrowding at Chartwell and that an extra cob was making trouble on the swan pool. We wrote and asked if we might borrow him, and Sir Winston replied that we could have him 'until such time as I may need him again'. He is a splendid bird and must have sired more than thirty cygnets since he has been at Slimbridge.

In Britain, the species is not yet considered to have established a self-sustaining breeding population so, unlike other introduced non-native wildfowl such as the Canada goose or mandarin duck, it is not yet on the official British List. Yet the black swan's original appearance here dates back as far as 1791, and the species first bred successfully in the UK, at Carshalton in Surrey, as early as 1851.

The most recent BTO *Bird Atlas*, using survey information from the years 2007–11, recorded black swans breeding – or attempting to breed – in more than 100 10-kilometre squares, while the species was present, but not yet breeding, in a further 150 squares. Black swans now regularly breed on the Norfolk Broads, while the Devon seaside resort of Dawlish is so proud of its resident birds that the species has been adopted as the town's emblem. Given that they appear to be able to cope with the vagaries of the British climate, and have no enemies apart from human beings and dogs, it may only be a matter of time before the black swan is adopted as a genuinely British wild bird.

Most people are now familiar with the black swan, if not in the wild, then through zoos and bird collections. But until just over 300 years ago, no one outside the original inhabitants of Australia had ever set eyes on one.

That all changed when, in January 1697 – more than seventy years before Captain Cook made landfall on this vast continent – a Dutch sea captain named Willem de Vlamingh came across flocks of black swans while venturing upriver from the western coast, along what is now the Swan River in the city of Perth. It is impossible for us to imagine the reaction of De Vlamingh and his crew to seeing what were clearly swans, but which, instead of being the familiar white colour, were jet-black.

When the expedition finally returned to Amsterdam in August 1698, the explorers brought back drawings of black swans; they had tried to bring live birds, too, but unfortunately the birds died on the long voyage home. De Vlamingh also died, either during the journey or very soon afterwards, and not surprisingly, in the absence of actual specimens, people began to doubt whether black swans existed at all. Even the scientific name first given to the new species, *Anas atrata* (meaning 'duck clothed in mourning'), sowed the seeds of doubt, suggesting that the bird discovered by De Vlamingh was perhaps a type of large duck or goose.

As Ian Fraser and Jeannie Gray note in their book *Australian Bird Names*, the species' very existence posed 'a serious challenge for Northern Hemisphere traditionalists, for whom "white swan" was a tautology and "black swan" an oxymoron!' But as Europeans finally began a permanent colonisation of Australia, starting with Captain Cook's landing in April 1770 and continuing with the arrival of the First Fleet in 1788, it soon became clear that black swans not only *could*, but actually *did*, exist.

Perhaps the scepticism was because 'black swan' was already a concept widely used in the English language to indicate something proverbially rare or non-existent. For example, in 1694 – just three years before De Vlamingh came across the species – *The Ladies Dictionary* commented on 'husbands without faults (if such black Swans there be . . .)'. This concept goes all the way back to Ancient Rome: the poet Juvenal, at the turn of the second century AD, wrote that 'a good person is as rare as a black swan'.

The longstanding denial of the existence of black swans epitomises the way in which human beings tend to view the world: if we have never seen something with our own eyes, we often assume that it simply cannot occur. As the Victorian philosopher John Stuart Mill points out, using this very example, absence of evidence is clearly not evidence of absence: 'To Europeans, not many years ago, the proposition that all swans are white, appeared an . . . unequivocal instance of uniformity in the course of nature. Further experience has proved . . . that they were mistaken.'

In other words, as soon as the birds were not only seen by reliable witnesses, but actually brought back to Europe to confound the sceptics, the entire argument collapsed. It is a classic instance of what a later philosopher, Karl Popper, defines as the 'Falsification Principle': the idea that a scientific statement must, at least theoretically, be able to be contradicted by new evidence, and therefore can (again, in theory at least) be proved false.

To some readers, this may appear either self-evident, or an arcane philosophical point with no obvious application to the real world. But examples of disastrous and (mostly) unforeseen events – such as the economic crash of 2008, or the more recent Covid-19 pandemic – clearly show that otherwise intelligent people may choose to ignore the possibility of something catastrophic occurring, simply because it hasn't happened yet.

In 2007, a full year ahead of the banking crisis that devastated global economies and had an overwhelming impact on the lives of billions of people, and more than a decade before the coronavirus pandemic, the Lebanese-American academic and economist Nassim Nicholas Taleb published *The Black Swan: The Impact of the Highly Improbable*. In the book, he discusses both the existence of 'black swan events' and – more importantly – how to safeguard ourselves against them occurring in the future.

As Taleb points out, the very complex nature of global institutions such as trading firms and banks, and the Byzantine systems they have developed, make them increasingly vulnerable to unforeseen catastrophes. He also notes that US bankers, in particular, had become so used to years of rising profits and investment success that they had come to assume this was the norm. In reality, the more years that markets rose, the more likely it was that the crash would come soon – or to put it another way, the more white swans you see, the more you should expect to eventually see a black one.

A year later, the entire global financial system was on the brink of collapse: an event Taleb had indeed foreseen, even if he could not have predicted exactly *when* it might occur.

In one memorable passage, Taleb points out that not everyone sees the world in the same way: every Thanksgiving, the slaughter of millions of turkeys may come as a surprise to them, but not for the butchers doing the killing, or the families sitting down to a turkey dinner. Hence Taleb's comment that we should try to 'avoid being the turkey'. Or, to put it even more simply, he tells us that we need to 'turn the black swans white'.

The irony of all this is that, even after the success of Taleb's book, we then still failed to anticipate what could turn out to be the biggest and most catastrophic black swan event ever: the Covid-19 crisis. This, too, was not accepted as likely to happen – even though, with hindsight, it was clearly only a matter of time before a global pandemic would emerge.

In fact, because it had been widely suggested that a pandemic would eventually occur at some time in the near future, some commentators are now suggesting that the Covid-19 crisis was actually as a 'white swan event' – one that could, and should, have been predicted.

In a further bizarre twist to the story, scientists have recently mapped the genome of the black swan, the better to understand how its immune system responds to avian influenza (commonly known as 'bird flu') – a pathogen considered far more lethal to humans than Covid-19. So far, since bird flu first transferred from birds to humans in 2003, just 800 people have been infected – but the bad news is that more than half of them have subsequently died.

Of all bird species, the black swan appears to be the most susceptible to the disease – indeed, in November 2020 Dawlish's birds died from avian flu. As a result of its vulnerability, the species is now being closely studied, with scientists comparing its immune response to that of the mute swan, its closest relative. If they do succeed in working out why black swans are so vulnerable, it is even possible that the species could turn out to be our best weapon against what might be an even more lethal pandemic – potentially the next 'black swan event'.

*

Mention the phrase 'black swan' to cinemagoers, and the chances are they will recall the critically acclaimed 2010 film of that name, for which leading actor Natalie Portman won an Academy Award for her intense and passionate performance as Nina, a ballet dancer performing a production of Tchaikovsky's celebrated ballet *Swan Lake*.

One of the most popular ballets of all time, *Swan Lake* tells the story of a beautiful princess, Odette, who is transformed into a swan by an evil sorcerer, Von Rothbart. During the hours of daylight, she appears as the Swan Queen, swimming on a lake of her own tears, but by night she takes on human form again, and is wooed by the handsome Prince Siegfried. He must swear an oath of his undying love and promise to marry her – which will break the spell and allow her to return permanently to human form. When he does so, however, he realises he has been tricked into proposing to the sorcerer's daughter in disguise. In the ballet's dramatic finale, Siegfried and Odette throw themselves into the lake and drown.

In the film *Black Swan*, the plot has a clever twist: Nina is required to play two versions of the lead character: the pure and innocent 'White Swan', and the dark, sensuous and potentially evil 'Black Swan'. In a further plot twist, Nina has an intense rivalry with a younger dancer in the company – Lily, played by Mila Kunis – which causes her to spiral down into a manic form of insanity, illustrated by increasingly bizarre scenes which may be real, or just a product of her fevered imagination.

Before becoming an Oscar-winning actor in *Black Swan*, Natalie Portman had studied psychology at Harvard University; something she put to good use during the year-long, physically and mentally demanding preparations for her role. Nevertheless, the entertainment journalist Hannah Wigandt later reported that Portman suffered mental health issues during rehearsals and filming:

> The distinction between Portman and her character Nina Sayers was sometimes dim, especially when Portman was training to become the ballet dancer. Portman was essentially playing an artist

who was pressured into having unrealistic expectations of herself as she tried to be perfect, even as Portman herself had unrealistic expectations put on her to become the character.

Cygne noir

More disturbingly, the depiction of the black swan as the sinister 'evil twin' of the pure white variety clearly has its origin in the largely negative way we perceive almost all black birds (with the odd exception of the blackbird itself). This derives from a broader cultural trope: that 'black' is linked with night, darkness, plague (the Black Death) and so on; while 'white' symbolises purity, chastity and other noble virtues. Many cultures – ancient and modern – regard the various members of the crow family, including carrion crows, rooks and ravens, as evil, and it is perhaps no surprise that the film adopts this characterisation for the black swan.

Now that we are more aware of the implications of these negative associations, especially when they are transferred to people of colour, it may be that they will begin to fade; though given how strongly they appear to be rooted in our collective psyche, this may take a very long time.

Not all people regard black swans in a negative way, though. The ancient indigenous Australian culture known as 'Dreamtime' includes a story of how the white swan Guunyu was attacked by the other birds, which were envious of his beautiful plumage. They then ambushed him and tore out his feathers. A crow came to the swan's aid, staunching his bleeding wounds and giving him his own black feathers, turning Guunyu into a black swan with a blood-red bill. In 2020, the artist Cheryl Davidson used this traditional tale as the inspiration for her striking redesign of the jersey for the Sydney Swans Australian rules football club, featuring a stunning image of a black swan.

The world's largest species of swan – and the only one I have yet to see in the wild – is the trumpeter swan of North America, named after its distinctive call. At up to 1.8 metres (almost 6 feet) long, with a wingspan of 2.5 metres (over 8 feet), and weighing up to 12.5 kg (27.5 lbs), the trumpeter swan is a tall, statuesque and very impressive creature. It is the heaviest bird in the whole of North America, and the largest species of wildfowl in the world.

As its name suggests, the species has a deep, resonant, trumpeting call, which carries for long distances across the lakes and forests of the Pacific North-west. It is able to call so loudly because it has a much longer and straighter neck than the more curved one of the mute swan. The bird's windpipe (or trachea) is looped at the point where it meets the breastbone, while the bronchi – the airways that lead from the windpipe into the lungs – have a more complex structure than those of other members of its family.

The trumpeter swan shares its long neck and loud call with its close relative, the whooper swan of Eurasia, the only obvious difference between the two species being the trumpeter's all-black (rather than black-and-yellow) bill. Indeed, until the latter half of the twentieth century the two were often considered to be different races of the same species, whose range covered the whole of the northern regions of the Americas, Europe and Asia.

Perhaps it was the lack of certainty as to whether the trumpeter swan was a species in its own right that led to it nearly going extinct. Once, this huge swan was so common that the early-nineteenth-century bird artist John James Audubon described 'flock after flock coming from afar' to alight on an icebound river, but by 1932 the total world population was thought to be down to just sixty-nine individuals. All these were inside the boundaries of Yellowstone National Park, where they took advantage of the famous hot springs to stay put all year round.

The trumpeter swan had been driven to the brink of global extinction partly by shooting for food, and also through the loss of its wetland habitats, but mainly by the lucrative trade in swan skins and feathers. The feathers were widely used to make quill pens, and were also in great demand for use in ladies' fashions in North America and Europe. From 1820 to 1880 the Hudson's Bay Company was responsible for killing as many as 108,000 trumpeter swans, to supply the London fashion market alone.

Fortunately, the passing of the Migratory Bird Conservation Act in 1929 halted the decline and, at the eleventh hour, protected both the swans and their precious habitats.

Subsequently, in the 1950s, an aerial survey along the Copper River in a remote region of Alaska came across an isolated, and hitherto unknown, breeding population of several thousand trumpeter swans. Here, far to the north, they could take advantage of the long hours of summer daylight and abundant food. In autumn, they would head south to spend the winter months in milder climes.

Over the next few decades, thanks to careful protection, numbers have continued to rise. Today, there are an estimated 60,000–76,000 trumpeter swans in North America, and the species is considered safe from any imminent threat of extinction.

Its migration routes are still not fully understood, and so in 2020 swans in Minnesota were fitted with tracking devices to enable scientists to discover more about their movements. Yet despite the trumpeter swan's recent return from the edge of oblivion, it remains legal to hunt them: in just one state, North Dakota, 2,200 licences for shooting swans were issued in 2020, allowing the birds to be shot from October through to December. While this may be legal, it does seem rather distasteful and unnecessary.

For the past half-century or so, since it was founded in 1968, the Trumpeter Swan Society has dedicated itself to protecting the species, as its mission statement declares:

Our mission is to assure the vitality and welfare of wild trumpeter swans . . . Early settlers nearly destroyed this magnificent species by the late 1800s. However, over the past few decades, people have passionately and with immense dedication worked to put trumpeters on the journey to recovery. You can too. Your support ensures trumpeter swans will not disappear in our modern world.

Like other once rare, but now reasonably common, birds, the trumpeter swan could easily decline again as a result of sudden and catastrophic environmental change. So even now, it needs all the friends it can get.

2

SPRING

If you're looking for monogamy, you'd better marry a swan.

Nora Ephron, from
the movie *Heartburn* (1986)

Mute swans, famously, pair for life. It is one of the few well-known 'facts' about swans that actually turns out to be (well, mostly) true. And although you might think that lifelong monogamy would be usual in the bird world – after all, every spring we see birds pair up to breed – in fact it is the exception rather than the norm.

There are close to 250 different families of birds in the world, and of these only a handful – including albatrosses and petrels, eagles, cranes, some parrots and owls, zebra finches, geese and swans – habitually remain faithful to their mate for the whole of their life. The rest either pair up for a single breeding season, or are serially unfaithful, in order to maximise their chances of raising young, and so ensure that their genes pass on to the next generation.

So why *do* swans mostly stay together for life? The key reason is that – as with other large, long-lived birds such as albatrosses – their offspring require long-term commitment from both their parents in order to rear them to adulthood, as they stay with them for much longer than short-lived species. Mute swans are mainly sedentary, and so pairing for life is relatively straightforward, but even those swan species that migrate, such as whooper and Bewick's swans, do so in pairs or family groups, enabling them to stay together throughout the year.

And while 'divorce' in mute swans does occasionally occur, studies have shown that fewer than 6 per cent of pairs – roughly one in seventeen – separate. This sometimes happens, as Christopher Perrins points out, when two pairs have a territorial dispute, during which the avian equivalent of 'wife-swapping' occurs. Normally, however, only when one of the two dies will the remaining bird take another mate. Bewick's swans are even more faithful: of all the many thousands of pairs ever to visit the Wildfowl and Wetlands Trust at Slimbridge, only two have ever separated while both were still alive.

Like other long-lived birds, swans take plenty of time to attain full maturity, usually breeding for the first time when they reach the age of three or four. For mature pairs, preparation for breeding actually begins the previous autumn, when adult swans gradually become less and less caring, and ultimately actively hostile, towards their now full-sized offspring. Eventually, the young either leave of their own accord, or are forced out by their parents, loafing around in grassy fields and along waterways in large, untidy flocks, in what a friend of mine calls 'Club 18-30 for swans'.

With the previous year's young now off their hands, the adult pair can now prepare for the most crucial annual event in their lives: the breeding season. They need to find – and then defend – a territory, which can vary enormously in size. Along rivers, each pair of swans may 'own' a stretch of water between 2.5 and 3 kilometres (roughly 1.5–1.9 miles) long; on enclosed lakes or gravel pits the territory may be smaller, so long as the pair are not within sight of a rival couple. The exception to this rule is the famous Swannery at Abbotsbury in Dorset, where hundreds of pairs nest next side by side with one another, to take advantage of a reliable and plentiful source of food provided by humans (see Epilogue).

In early spring, once the pair have chosen where they will breed, courtship begins. Like everything else in a mute swan's life, this is a very public affair. As with all avian courtship displays, this is about creating a bond with one another, or, in the case of mature pairs who have already bred together before, re-creating that bond. The whole process can take several days, with the male beginning the ritual by following his mate around, until she finally responds by turning towards him and engaging in a mutual display.

Both birds fluff up their wing feathers and then pose face to face, each one mirroring the other's movements. They may turn their heads from side to side, dip their bills into the water, and then, at the climax of the display, lower their heads and touch their bills together, to form the classic 'heart-shape' image so beloved of the designers of Valentine's Day cards.

The mating of swans – as with other wildfowl species – is less romantic than this preamble might suggest. To a casual observer it can appear as if the male is trying simultaneously to have sex with and drown his mate, as he climbs onto her back, grabs her neck with his bill, and sometimes even forces her momentarily beneath the water's surface.

Mercifully, the actual copulation period lasts just a few seconds, after which the male releases the female, and they again engage in the delicate rituals of courtship. This process is usually repeated a number of times, both to ensure that the eggs are fertilised, and to cement the bond between the pair.

Soon afterwards – sometime between the beginning of March and the end of April, depending on the location and the prevailing weather conditions – they will start nest building. Like everything else about mute swans, the nest is huge: by far the largest of any British bird, though according to wildlife photographer Malcolm Schuyl it is not so much a nest, more 'a heap of vegetation'.

In his 1949 novel *Adventure Lit Their Star*, Kenneth Allsop includes this vivid description of the nest structure:

A pair of mute swans had a nest on the turf dike separating two lagoons, a huge pyramid of rushes and rotting cabbage leaves. Mounted on top like an heraldic emblem was the pen, or female, in the first stage of her five weeks' incubation of the eggs. Beside her on the dike was her mate, the cob.

Allsop then describes the male's reaction to the unexpected arrival of a pair of little ringed plovers near the nest (following which, perhaps not surprisingly, the plovers immediately flee):

It watched the arrival of the plovers with suspicion in its squinting pig-eyes. When it saw that they had the temerity to land only twenty yards from the nest, it overrode the hissing of the pen with a trumpeting grunt and waddled upon them with great wings arched and neck coiled back.

The choice of where to nest is initially made by the male, who also does all the construction work. Patiently and methodically, he seeks out and brings back a varied assortment of living and dead water plants, including reeds, rushes and waterweed, along with other material such as sticks and grasses. He then assembles this into an untidy structure up to 2 metres across and 75 cm high (6.5 by 2.5 feet), with a shallow depression in the centre where the female will lay her eggs.

Even when a male has almost completed the nest, his mate, having inspected it, may decide that it is not suitable, forcing him to start building all over again nearby. The 'best' place for a nest is inevitably a compromise: it needs to be close to the water, so that the pair can feed easily, but not in an area that might flood (which is why nests along rivers are often built on dry land near the edge of the water, to safeguard against sudden and unexpected rises in water level).

There also has to be an abundant and easily accessible supply of food – mainly aquatic vegetation such as waterweed and algae – while the area around the nest site should, if at all possible, be undisturbed. Disturbance and danger, as we shall discover, can come from human beings and their dogs, as well as from other river creatures like otters.

Even so, mute swans regularly build their nest slap bang in the middle of, or very close to, a human thoroughfare, such as a canal or river towpath, as the ornithologist Richard Fitter noted in his 1949 book *London's Birds*:

> Mute swans . . . make their bulky piles of weed and other rotten vegetation on the ground . . . sometimes right next to a public path, within a stone's-throw of the water. Being large birds, awkward on land, they like to have as little walking to do as possible, and are sufficiently fearsome to scare human marauders away.

They do not always succeed: as Fitter goes on to reveal, in spring 1947 two swan's eggs were stolen from a nest by the Long Water in Kensington Gardens, a crime that Fitter points out is 'hard to disassociate from the small size of our post-war egg ration!' Urban swans will also build their nests out of any material to hand: one nest on London's Regent's Canal was almost entirely woven out of electrical flex and bits of rope.

Where I live, on the low-lying Somerset Moors and Levels, the land is criss-crossed by rhynes – narrow, steep-sided, water-filled ditches – which are usually just broad and deep enough to support a pair of swans. Walking or cycling around from late March through to May or June, I frequently come across swans' nests wedged against the bank and often surrounded by a thick layer of apple-green duckweed, almost covering the water's surface.

I'm surprised that the rhynes are so popular: the banks are high enough to limit the view of the incubating female; nevertheless, I rarely come across nests that have been found by the local fox or other predators. Maybe the male, who usually stands guard on the bank above his mate, poses enough of a deterrent.

Once the female has agreed to a site, and the nest is ready, her next task is to lay a clutch of eggs. She does this sometime between late March and early May, with a mean date of 18 April. Producing the clutch is no small feat: typically, mute swans lay between five and seven eggs, but up to twelve have been recorded. Unusually for birds, the eggs are not wider at one end than the other, but symmetrical ellipses – effectively a circle stretched into an oval.

Incidentally, the later the first egg is laid in the season, the smaller the total clutch is likely to be. This may be because later nesters are less experienced birds, perhaps breeding for the very first time; or because the later clutches are repeat ones, laid after the first has been lost or destroyed.

The eggs themselves are off-white in shade, with a tinge of bluish-green or grey, giving them a rather 'dirty' appearance, which increases over time as they are gradually stained by the damp vegetation lining the nest. They are roughly 11–12 cm (4.3–4.7 inches) long by 7 cm (2.75 inches) wide, and weigh about 350 grams (just over 12 ounces) – five or six times that of a large hen's egg – though the shell is just 1 mm thick.

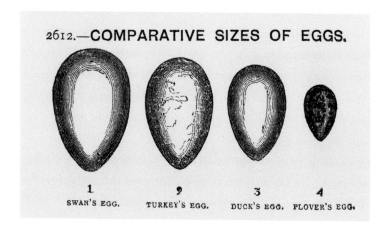

2612.—COMPARATIVE SIZES OF EGGS.

1.	2	3	4
SWAN'S EGG.	TURKEY'S EGG.	DUCK'S EGG.	PLOVER'S EGG.

Though a swan's eggs are the largest of any British bird, an average clutch of six eggs (weighing about 2.1 kg) represents only about one-quarter of the female's body weight. In contrast, a standard clutch of eggs from Britain's smallest bird, the gold-crest, weighs over 5 grams – about the same as the bird itself.

A female mute swan lays an egg roughly every two days. Once she decides that the clutch is complete, she will begin to incubate, ensuring that the cygnets will all hatch out at more or less the same time. Throughout the incubation period, which lasts for five or six weeks, she will sit more or less solidly, keeping the eggs warm by covering them with the bare brood patch on her belly. For the whole of this time she hardly feeds at all, instead relying on her stored reserves of protein, fat and minerals. As a result, she may lose up to one-third of her body weight.

During the breeding season, male swans in particular can be very antagonistic towards any rival swan that dares to venture into its territory. Intruders – whether they have arrived intentionally or not – are greeted with immediate and powerful retaliation. In a display known as 'busking', the incumbent male will approach the interloper, raising his wings to assume the largest and most threatening posture he can manage, while erecting the feathers on his neck and advancing towards it by paddling both feet in unison.

Once he reaches his rival, they engage in battle: intertwining their necks and hitting one another with those powerful wings. The aim of the incumbent is either to see off his rival or drown it – and, occasionally, he succeeds in doing so.

In 2018 reports emerged of a male mute swan at Castle Pond in the Welsh city of Pembroke that had attacked and killed at least ten – and possibly as many as twenty – other swans. These birds had inadvertently trespassed into his breeding territory when they swam over a sluice and into the pond, from which, because of the steep stone walls surrounding it, they were then unable to escape. The enraged swan – inevitably dubbed 'Mr Nasty' – would often climb on top of the intruder and drown it. This reminded locals of a similar swan from a few years earlier which also attacked any swan that strayed into the same pond. His nickname? Hannibal.

Swans also have a reputation for belligerent behaviour towards other waterbirds. But might they actually be even more aggressive towards members of their own species? A study carried out by scientists from the Wildfowl and Wetlands Trust and the University of Exeter suggests that they are.

Using observations made by students, observing footage from webcams at the WWT centres at Slimbridge and Caerlaverock, the scientists collated examples of interactions between various waterbirds, and discovered that swans do tend to focus their hostility more towards their own species than to others. Given that they compete with one another for food and other resources such as nest sites, this is perhaps hardly surprising. What was, however, more unexpected was that the species that showed the greatest tendency to infighting was the smaller, and gentler-looking, Bewick's swan.

*

Ironically, for a bird that is so well loved by the general public, almost all threats to nesting swans come from humans: either directly, in deliberate or accidental disturbance, or indirectly, through wayward dogs and, on rivers, the wash from passing boats, which may flood the nest and eggs. During the initial lockdown due to Covid-19 in spring 2020, swans and other waterbirds nesting along the banks of the River Thames had their most successful breeding season for years, as boat traffic virtually disappeared.

Swans are always wary of close approach by human beings, and will habitually hiss at anyone who gets too near. But during the breeding season, and especially when the female is incubating her precious clutch, this behaviour is heightened, as Thomas Bewick points out in the 1804 volume of *A History of British Birds*:

> While they are employed with the cares of the young brood, it is not safe to approach near them, for they will fly upon any stranger, whom they often beat to the ground with repeated blows; and they have been known by a stroke of the wing to break a man's leg.

The widely held – and totally erroneous – belief that a swan can break a man's leg (or arm), with a single blow of his wing, shows no sign of going away. The Victorian nature writer Edward Stanley, Bishop of Norwich – who appears to have believed almost any nonsense about birds – hedges his bets on whether or not it is true:

It is a prevailing opinion, amounting almost to a proverb, that a stroke of a Swan's wing will break a man's leg. How far this may be strictly true we cannot say; but . . . we think it not impossible.

My old friend and colleague Bill Oddie once told me that when he, Tim Brooke-Taylor and Graeme Garden were on location filming *The Goodies*, they were often approached by what he called 'the town bore'. To amuse themselves, the trio would turn the conversation around to swans, and then time how long it took their interlocutor to utter the immortal phrase 'Swans! Break a man's arm, swans can!' The record was, apparently, three seconds.

It goes without saying that any swan that attacks a human intruder is at far greater risk of breaking its own wing, as its bones need to be light and delicate for the bird to get airborne. If that is the case, then why do so many male swans risk their lives by doing so? Partly this is out of necessity: the swan has no idea whether a human being approaching their nest has benign or hostile intentions – and, as I discovered when researching this book, many do fall into the latter category. Either out of ignorance or malice, or perhaps a combination of both, many people do appear to have it in for swans.

One group of people who have been unfairly singled out as 'swan killers' are Polish immigrants to the United Kingdom. This began after Poland gained full membership of the European Union in April 2004, which allowed free access to all other EU member countries, including the UK.

Over the next decade or so more than one million Poles came here to live and work, making a largely positive contribution to the nation's economy. However, their arrival and presence – especially in some smaller, rural communities – inevitably generated a xenophobic backlash, with negative stories about them circulating in the media. Of these, perhaps the most prominent and persistent is the idea that 'Polish people eat swans'. Several newspapers, including the *Sun*, the *Star* and the *Daily Mail*, ran stories with lurid headlines claiming that 'the Queen's swans' were being slaughtered and eaten by Polish immigrants.

This soon developed into a fully fledged urban myth – defined as 'a story or statement that is not true but is often repeated, and believed by many to be true'. Rather like the notion that a swan can break a man's arm, we are now saddled with the idea that Poles routinely kill and eat swans, even though there is no evidence for it actually happening. Such is the power of prejudice expressed through the popular press.

Now, the Poles are fighting back. On one YouTube video, entitled 'We eat Swans?! Tweets about Polish people #14', a young Polish woman ('Marticore') roundly debunks a range of lazy assumptions about Poles using a healthy dose of humour and sarcasm: 'Who the hell would eat swans?! No, we don't eat swans, English people – your holy birds are safe!'

Ironically, the worst ever recent massacre of swans in the UK, which took place a mile or so down the road from my Somerset home in early 2011, was carried out by neither Poles nor reckless youngsters. The likely culprits, a group of local, middle-aged men, who shot first eight and then a further twenty-three mute swans in the head with airguns, appear to have got away with their horrific and pointless crime, as there was not enough evidence to convict anyone. This was despite the offer of a substantial reward: initially £10,500, and later raised to over £25,000, thanks in part to a generous donation by the author Terry Pratchett. The money did not go to waste: the local animal rescue charity Secret World, which had organised the reward fund, pledged to use the cash to help protect swans in the future.

In April 2009, another intriguing story appeared in the tabloid press. A mute swan – who, in the long tradition of such names, was dubbed 'Mr Asbo' (after the acronym for the government's Antisocial Behaviour Orders, used to control a range of minor offences) – was terrorising rowers on the River Cam, just outside Cambridge. From time to time, his attacks were so persistent that the boats actually capsized.

Predictably, perhaps, the rival camps took militantly opposing views on what should be done with the swan, which, after all, was only defending his riverine territory against what he regarded as intruders. While the RSPCA defended his behaviour as entirely natural, the Conservators of the River Cam – the body which officially manages the river, and its human and animal occupants – decreed that the swan's behaviour had gone 'out of control'. The situation was not helped by some newspapers, who allegedly paid rowers to approach the bird so they could get more dramatic photos of the attacks.

Over the next few years, various attempts were made to solve the problem: initially by moving him a mile or two upriver, though of course he always found his way back. A local psychic (inevitably described as 'the swan whisperer') even claimed she could persuade Mr Asbo to mend his violent ways with a quick woman-to-bird chat. Unsurprisingly, this didn't work either.

Finally, in May 2012 – more than three years after the saga had begun – Mr Asbo and his mate were captured, had their wings clipped, and were moved to a swan sanctuary 60 miles away. This was, perhaps, the best – or least bad – solution, given that by then the swan was attacking not just rowing boats and punts, but also larger vessels with outboard motors, causing real concern that he might injure or even kill himself.

But the story was not quite over. A year before Mr Asbo and his mate were removed, his son – inevitably named 'Mr Asboy' – had followed in his father's footsteps (wingbeats?) by attacking rowers and tourists on punts. And in spring 2019, his grandson 'Asbaby' did the same. I say 'grandson', but there is no evidence that the birds are related at all.

In the *Guardian*, Asbaby featured in the newspaper's 'Pass Notes' column, which takes a wry look at the day's news. 'Have generations of an entire flock been driven mad by posh students in rowing blazers?' it asked, rhetorically, though it went on to point out that in 2012, in the US state of Illinois, a man allegedly drowned after a swan tipped him out of his kayak and prevented him from swimming to safety. This, incidentally, may also be an urban myth.

★

Given the fuss made about a single swan terrorising river-goers on the Cam, you might assume that such conflicts are typical. In fact, although male swans will often defend their territory with threatening postures and sounds, they rarely resort to actual physical attack. A valid comparison might be made with the way we demonise sharks for their (admittedly more horrific, but globally very infrequent) attacks on humans.

Just as sharks suffer far more from us than we do from them, so do swans. Researching this book, I came across dozens of cases of swans either being deliberately attacked or accidentally harmed by people and their dogs. These range from horrific acts of malice to simple carelessness, but the consequences are often the same: the permanent injury or death of the swan.

In June 2020, for example, a dogwalker came across a dead mother swan and the corpses of three of her seven cygnets along the River Thames at Waterhay, near Cricklade in Wiltshire. These had been killed by three youths using a powerful catapult, firing ball bearings at the unfortunate swans. Despite the offer of a £500 reward, no-one has been arrested or charged with this crime.

Horrific though that incident was, when it comes to swan killing, it is merely the tip of a very large iceberg. In Kent in 2018, an entire family of seven swans – the cob, pen and five cygnets – was killed after being peppered with shots from airguns; X-rays revealed dozens of lethal pellets inside their bodies. In July 2020, the bloodied body of a cygnet was found on a golf course at Hetton-le-Hill, near Sunderland. The same month, a swan with a crossbow bolt in its neck was photographed near Thrapston in

Northamptonshire; while one more fortunate swan was rescued after being shot in the head at Newbury in Berkshire.

During the lockdown in spring 2020, many people connected – or reconnected – with the natural world on their doorstep. Yet the man in charge of the Queen's swans, Royal Swan Marker David Barber, suggested that the number and severity of swan shootings spiralled during that period, and was by far the worst he had known in almost thirty years in the post. He blamed a combination of 'bored and irresponsible' airgun owners and the fact that nesting swans are so easy to shoot – literally a sitting target. It should also be pointed out that we do come across dead swans more often than the corpses of other birds, as their size and colour make them a lot more obvious.

Although airguns and air rifles are cited in the majority of re-ports, some incidents involve boats being deliberately aimed at swans. In July 2020, a motor boat was driven at speed at a swan in Langstone Harbour, Hampshire, even changing course to target the bird. Swans are vulnerable on land, too: a cygnet was mown down by a passing motorist in Ulverston, Cumbria; a bird rescue volunteer arrived to find the distressing scene of the rest of the family gathered around the bloodied corpse of their dead sibling.

Not all injuries and deaths are caused deliberately. Discard-ed fishing tackle, string and various items of litter can all cause swans to become entangled, and sometimes die as a result – es-pecially if the material becomes entwined around their beak, preventing them from feeding.

After face masks became compulsory during the Covid-19 pan-demic, discarded masks soon became a huge problem for swans and other wildlife. However, one swan in a French park got its own back by pecking aggressively at a face mask of a woman who got just a bit too close for the bird's comfort. The resulting brief video of the incident went viral, with more than 28 million views. Many people pointed out that the swan had managed to pull the mask, which the woman had been wearing around her neck, back over her mouth and nose, where it should have been in the first place.

Dogs, too, can wreak havoc among nesting swans. Cygnets are especially vulnerable, but adult swans are also often injured or killed by dogs. And in Derbyshire, five cygnets were left orphaned when a dog killed one of their parents. The situation is not helped by the fact that in England – unlike Scotland – the owner of a dog that attacks a swan is not deemed legally responsible, and so cannot be prosecuted.

Sometimes a bird attacked by a dog is found alive, but too badly hurt to save; as happened in Warrington, Cheshire, when a swan was found with terrible injuries to its wings. Despite the best efforts of the local RSPCA, it had to be put down. The man who found the swan, Paul Bray – himself a dog owner – commented that 'Yet another innocent, magnificent, proud, feisty swan has been lost because of the selfishness of humans.' Two weeks later, in another part of Warrington, another swan was also found horrifically injured, and again had to be humanely killed.

Occasionally, swans take revenge on their canine adversaries. In July 2019, while swimming in a Dublin park lake, a cocker spaniel got rather too close to a family group of swans and their cygnets. Moments later, after two or three rapid blows of the adult swan's wing, the dog disappeared beneath the surface of the water and drowned.

To try to prevent any harm coming to either swans or dogs, concerned residents on the Isle of Wight erected a sign asking dog owners to keep their pets on a lead while passing a site where swans are regularly fed. Predictably, the sign was at first vandalised and then removed.

What is so astonishing about the number of incidents I have described is that they – and countless others in all parts of the United Kingdom – almost all took place in just *one* calendar month: July 2020. That was, of course, the time that lockdown restrictions were being relaxed, so more people would have ventured outdoors; even so, the frequency is telling.

There are, occasionally, positive outcomes to these terrible attacks. In August 2020, the Berkshire-based charity Swan Support reunited an injured cob swan with its family at Beale Park, alongside the River Thames near Pangbourne. The male had been shot in the neck earlier in the year, but after it was taken into care it eventually recovered. This was one of three separate incidents of mute swans being shot there during that spring and summer.

Meanwhile, at Roundhay Park in Leeds, during the first lockdown period one young boy became fascinated by the local swans. Ten-year-old Dylan Tripp had been very upset when first a nesting swan was savaged by a dog, and then he and his father Richard witnessed a dog owner throwing a stick near another nest, to the great distress of the cygnets. Following the incident, Dylan wrote a delightful story about the swans, including this vivid and timely warning to dog owners to take more care of their pets:

Man's best friend is a friend to you and me but not swans; they despise them. If a dog does come too close to the swans they will hiss and hiss and hiss and translated that is 'go away I don't like you'. Most people only let their dogs in the water if there aren't any swans, ducks or geese near them but that's most people – not everyone. While they were following the path on a small bend a large hound came off its lead into the shallow water, the swans started to hiss but the dog stayed in the water getting closer. The cygnets are now learning how bad dogs are. The hissing continues and the adult swans are now violently shaking and puffing their wings and feathers out to get the dog away. The dog then goes away but the swans are left shaken.

With swans in jeopardy on so many fronts, it's not surprising that over the years a number of rescue centres have been set up. Of these, the best known is the Swan Sanctuary, situated on the River Thames in my home village of Shepperton. Dedicated to the treatment, care and, where possible, rehabilitation and release back into the wild of sick and injured swans and other wildfowl, it also plays an important role in training and education in the care of these birds.

The sanctuary began in humble circumstances back in 1979, when Dorothy Beeson found a swan entangled in fishing wire near her home in Egham, Surrey. She nursed the bird back to full health in her garden, and then began taking in other injured swans.

Soon, however, she had too many birds to cope with at home, so she sold her house and rented a 2-acre site on a derelict allotment nearby, where she and her partner Stephen Knight lived in a static caravan for sixteen years. Eventually, the council needed the land back, so in 2005 the sanctuary moved to a permanent site by the Thames at Shepperton, where it remains to this day. Staff and volunteers are on call twenty-four hours a day, 365 days a year to send out a rescue squad in a veterinary ambulance to an injured bird. Once the birds have recovered, they are either released back into the wild or, if they are not able to survive on their own, are rehomed elsewhere.

Like so many animal welfare charities, the Swan Sanctuary depends entirely on donations of cash from its many private and corporate supporters, and in kind – local supermarkets donate their waste food to feed the birds. Their list of patrons and former patrons includes Queen Noor of Jordan and the actor Michael Caine, who has visited the sanctuary, where he was photographed holding an injured swan.

Dorothy and her team have achieved public recognition for their work over the years: she was awarded the British Empire Medal (BEM) in 1991, followed by an upgrade to an MBE in 2015, 'for services to swan rescue and conservation'. The sanctuary itself has also won awards from the RSPCA and the International Fund for Animal Welfare, as well as the Queen's Award for Voluntary Service in 2018, presented by HRH Princess Alexandra. It has featured regularly on television, on prime-time programmes such as *Wildlife on One*, *Animal Hospital* and *Pet Rescue*.

Dorothy – always known as Dot – died after a long illness in May 2020, aged seventy-two. Reading her obituary, many of her supporters were surprised to find that her kindness went beyond her beloved wildfowl: for many years, she had also helped vulnerable children through the youth justice system. But she will be best remembered for her dedication to the welfare of our largest bird. The day-to-day running of the sanctuary is now taken care of by Dot's partner Stephen and her daughter Melanie.

*

Swans can find themselves in danger for many other, less deliberately malign reasons. They occasionally bring road traffic to a standstill by straying onto a motorway or dual carriageway, or cause chaos on our railway system when they land on the tracks. Such birds are often described in the media as 'rogue', 'pesky' or 'unruly' swans, as if they had deliberately chosen to behave delinquently to cause gridlock, or to have become lost or confused.

But one story from Germany firmly refutes this notion. Just before Christmas 2020, an adult mute swan was discovered sitting on high-speed railway tracks near Fuldatal in Germany. Initially it would not budge, causing at least twenty trains to be cancelled, until eventually it was captured and released on a nearby river.

It turned out that the reason the swan chose that particular spot was that its mate had recently flown into the overhead power lines above the tracks and been killed. So, rather than being confused, or having strayed accidentally onto the railway line, the swan was effectively mourning its partner. This was another confirmation, if one were needed, of the legendary fidelity of swans.

On occasion, swans exhibit devotion towards human beings as well. In January 2021, during the Covid-19 lockdown, Sophie Byers was working at home, in the village of Askam in Cumbria, when she noticed an adult mute swan walking past her window. Later on, she found it sitting right outside her front door. Wondering if the bird was lost, she spent an hour coaxing it back to the local pond; but as she walked home, she soon realised that the swan was following her back again – as she said, 'It did feel like I was being stalked.'

Wondering if the bird was ill, Sophie called the RSPCA, who in turn contacted Caroline Sim – known locally as 'the swan lady' – who came to pick up the bird, took him home and fed him, as he seemed very light for his size. After an overnight stay, having checked for symptoms of avian flu, she released the swan elsewhere. As Sophie reflected, 'We have a saying around here – "Only in Askam" – and this is definitely one of those times!'

Across the Irish Sea in Wexford, another swan took to 'hugging' cars, resting its head and neck on the still-warm bonnets of parked vehicles. It turned out that this swan had recently lost its mate, so perhaps was seeking comfort in the warmth of the engines, as it would previously have done by resting on its partner. And in an even more bizarre incident, in March 2021, an adult mute swan was reported to be using its beak to rattle letterboxes on a residential street in Northampton, which it had apparently been doing for the past five years. This may have been, as the RSPB suggested, because it was searching for food or more likely, given that it seems to happen in the early spring, marking its territory.

But all these brief incidents are trumped by the extraordinary story of one Turkish man, and his dedication to a single whooper swan he found with an injured left wing, back in 1984. At the time, Recep Mirzan was doing his rounds as a postman when he came across the bird lying in the middle of a field outside a village. Mirzan rescued the swan, which he named Garip, and nursed her back to full health.

But the story does not end there. Even though she was able to fly free, Garip stayed with Mirzan, walking around with him and keeping him company after his wife died. Almost four decades since he first came across her, she still does.

Recep Mirzan's close bond with his swan has echoes of a similar pairing more than eight centuries ago. During the twelfth century, Hugh, the bishop of Lincoln, also befriended a swan, which would follow him around as he undertook his holy duties. The swan became so devoted to his human companion that it would attack anyone else who approached him, which must have made his pastoral duties tricky at times. Soon after he died – with his faithful swan by his bedside – Hugh was canonised, and is now the patron saint of sick children, shoemakers and, of course, swans.

If a pair of swans manage to survive danger for the whole of the long (and, for the female, tedious) incubation period, the day finally arrives when the eggs begin to hatch. Like other species of wildfowl such as ducks and geese, the offspring of swans – known of course as cygnets, a term deriving via Anglo-Norman from the Latin word *cygnus*, meaning swan – are termed 'precocial' or 'nidifugous'. This simply means that they are capable of leaving the nest and finding their own food soon after hatching – unlike songbirds, which are born blind, naked and helpless, and entirely dependent on their parents for food.

The process of hatching is far from instantaneous: the long and arduous process begins at least forty-eight hours beforehand, when the chick starts to use a small, sharp protuberance on the end of its bill – known as the 'egg-tooth' – to force its way out of the shell.

This is a long, hard slog, especially for a tiny bird whose only experience of the world has been, until then, inside the confines of the egg. But, driven on by instinct, and helped by the gradual thinning of the eggshell towards the end of the incubation period, the chick persists, starting with small cracks and holes and eventually breaking open the entire shell to arrive in the outside world.

Even though the female does not start incubating until all the eggs have been laid, the cygnets nevertheless hatch out at different times. It usually takes twenty-four hours or so before the brood is complete, while some eggs may never hatch.

Cygnets are the proverbial 'ugly duckling': grey, rather than white like their parents, presumably to be less conspicuous while they are still so small and vulnerable. Yet just a day or two after emerging from the egg they will embark on the first of many adventures, usually accompanied by their father, while their mother remains on the nest to safeguard any cygnets that have hatched out later.

Like other precocial birds, including goslings and ducklings, cygnets (which oddly, are rarely known as 'swanlings') are able to swim, feed and generally look after themselves from a very young age. They may still be small – at birth they weigh about 220 grams (about 8 ounces) – but they are endlessly curious about their new surroundings. I have sometimes seen a female swan calling frantically to her offspring, as one or two always seem to be either lagging behind or heading out in front of the rest of their siblings, eager to explore their new and fascinating, yet potentially dangerous, world.

At this early stage in their life they are, of course, very vulnerable. Pike are one danger, as they can swiftly drag a cygnet beneath the surface from below; mink, otters, foxes, rats and herons are also partial to a fluffy snack. Many times, I have seen a family of seven or eight cygnets swiftly reduced to two or three, sometimes in just a few days. Other times, the whole brood manages to survive to reach maturity.

Cygnets will usually remain with both parents for between four and six months after hatching, and although they are able to find much of their own food, their small size means that the parents are often on hand to help them – for instance by stirring up the food so that it floats on the surface, where the cygnets can easily pick it off. Once they have left the nest they often travel some distance away: here on the Somerset Levels I often see swan families a mile or more from their original nest.

The young do grow very quickly: by September or October the early broods have usually reached adult size. But they take a surprisingly long time before they can fly: anything between four and five months, compared to just one and a half to two months for Bewick's and whooper swans. That, of course, is because those species breed at much higher latitudes, and so need to migrate as soon as the brief Arctic summer is over; as a result, the young must grow more rapidly than their more southerly, and sedentary, cousins.

Mention the phrase 'ugly duckling' to anyone of a certain age, and they are likely to start singing, or perhaps whistling, the Danny Kaye song of that name. Composed by Frank Loesser, who also wrote *Guys and Dolls*, the song features in the 1952 Hollywood musical *Hans Christian Andersen*, the biopic of the famous nineteenth-century Danish children's author.

Andersen's 1843 story, '*Den Grimme Aelling*', was translated into English three years later as 'The Ugly Duckling', and since then has been a favourite fairy tale for many generations of children. Yet like so many of Andersen's fairy tales, including 'The Snow Queen', 'The Princess and the Pea' and 'The Emperor's New Clothes', the story of 'The Ugly Duckling' is darker, and contains a more complex moral message, than might first appear.

The plot is simple: one of the birds that hatches out of an

egg in a duck's nest is abused by his fellow farm creatures for his dingy grey colour, and ends up as an outcast.

One day he sees a flock of wild swans heading overhead, but is unable to fly, so cannot accompany them on their travels. He hides away in shame for the whole winter; but when, in spring, the swans return, he cannot resist joining them. To his surprise and delight, they welcome him; only when he sees his reflection in the water does he discover why: he has been transformed into a pure white swan. 'His own image', writes Andersen:

> no longer a dark, grey bird, ugly and disagreeable to look at, but a graceful and beautiful swan. To be born in a duck's nest, in a farmyard, is of no consequence to a bird, if it is hatched from a swan's egg.

The tale is a classic example of Andersen's modus operandi: what his biographer Jackie Wullschläger calls 'presenting lessons of virtue and resilience in the face of adversity', not just for children, but their parents as well. The power of the story may derive from the fact that Andersen was, as a boy, tall, gawky and rather unprepossessing in appearance, and was teased unmercifully by his schoolmates.

Since the story first appeared, the metaphor of 'the ugly duckling' – defined by the *Oxford English Dictionary* as 'a young person who shows no promise of the beauty, success, etc., that will come with maturity' – has entered the English language. It appears in the writings of, among others, George Eliot (Maggie Tulliver in *Mill on the Floss*) and George Bernard Shaw (Eliza Doolittle in *Pygmalion*). The popular book-sharing website Goodreads even has a genre of romances labelled 'Ugly Duckling or Eliza Doolittle books'.

More recently, non-fiction writers have employed the image, too. In her 1963 work *The Feminine Mystique*, feminist Betty Friedan writes of 'the unhappy ones, the ugly ducklings, [forced] to find themselves while the girls who fitted the image became adjusted "happy" housewives'. Author and philosopher Criss Jami adapts the message of the fairy tale for a more straightforwardly stoicist homily, to champion those who do not undergo the final transformation, and are fated to remain 'ugly ducklings':

> There's no need to curse God if you're an ugly duckling. He chooses those strong enough to endure it so that they can guide others who've felt the same.

The concept, it seems, is reinvented to suit the author's particular message. Another consequence of the story's worldwide currency is that in 1984, following a vote on a popular television show, the mute swan displaced the skylark as Denmark's national bird.

Oddly, some mute swan cygnets are not grey at all, but pure white – just like their parents. Known as 'Polish swans' – because of a longstanding and erroneous belief that they originally came from Poland – they were hailed by the Victorian ornithologist William Yarrell as a new species, which he proudly christened *Cygnus immutabilis* ('changeless swan'):

> The London dealers in birds have long been in the habit of receiving from the Baltic a large Swan, which they distinguish by the name of the Polish Swan. I had reason to believe that this Swan would prove to be a distinct species, though even more allied externally to our Mute Swan, than the Bewick's is to the Hooper [sic] . . .

Yarrell goes on to relate the events of January 1838, when, during severe weather, several flocks of these swans were observed flying southwards along the east coast, from Scotland to the mouth of the Thames.

However, Yarrell's confident declaration of a new species of swan proved unfounded, with the bird soon relegated to being a race or subspecies of the mute swan. Today, we know that it is actually a colour morph – similar to light phase and dark phase

Arctic skuas – in which a single genetic mutation produces a recessive gene which, when both parents possess it, produces these unusually white offspring. They also have pale, pinkish bills and paler grey legs and feet which, like the distinctive white plumage, are a result of a reduction in the dark pigment melanin.

Having observed Polish swans at various sites in Norfolk from 1980 onwards, the ornithologist Moss Taylor investigated their origin, unearthing some fascinating details about their lives. Not least, he discovered that the all-white plumage can be a major drawback for the cygnets, which are not only more conspicuous to predators than their grey counterparts, but may also be driven away by adults because their white colour marks them out as a potential rival.

He also suggests that the lack of melanin in their bodies may result in weaker legs and feet, making them more prone to suffer injury. Their all-white plumage did, however, make Polish cygnets popular among clothing manufacturers in the Netherlands, who during the nineteenth century selectively bred these birds to produce white down to make children's slippers.

As soon as the cygnets have left their nest, they begin to feed. Both adult and young swans are almost exclusively vegetarian, though they do occasionally take aquatic molluscs and even very small fish, mostly by accident, when these are attached to their plant food. The edges of the swan's bill are serrated, enabling them to grip onto vegetation while they feed.

Adult swans (and well-grown cygnets) dip their very long necks beneath the surface of a pond, lake, river or stream, or 'upend' their bodies into a vertical position, enabling them to consume a range of aquatic plants such as pondweed. On coastal marshes and estuaries, they also feed on a range of vegetable matter, including eelgrass, an aquatic flowering plant that grows in huge clumps in the intertidal zone between land and sea.

At times, swans have come into conflict with anglers, who have wrongly claimed that the birds are eating their precious fish. While swans might consume fish eggs or very small spawn by accident when feeding on aquatic plants, it is unlikely that this would make any significant difference to fish stocks.

Out of the water, mute swans regularly graze for food, using their powerful bill-tips, with those serrated edges, to snip off the tender ends of grass shoots and other herbs. They also occasionally feed on cereal crops – though not perhaps to the extent that

geese do. However, being so large and heavy, they not only consume more than other birds, but their broad feet can also cause
damage to newly grown crops, especially in late winter and early
spring, when these are just beginning to emerge.

It is widely known that small birds, with their light weight and
rapid metabolisms, need to eat between one-third and one-half
of their body weight every single day of their brief lives, just
to survive. You might imagine, given its much larger size and
weight, that a swan would not need to consume anything like
as much, relatively speaking. Yet the few studies that have been
undertaken suggest that an adult mute swan eats between 3.5
and 4 kg (7.7 to 8.8 lb) of 'wet vegetation' a day – equivalent to
about one-third of its body mass. The reason is that most of a
swan's diet is nutritionally poor, so, like cows, they must feed
more or less continuously to get enough energy from their food.
They also regularly consume grit, to aid their digestion.

Swans regularly take food from humans, especially by the
sides of park ponds and lakes. Unfortunately, the standard offer of white bread is very poor, nutrient-wise, and can cause the
birds to develop digestive problems. Seeds and grain are much
better for them, though given how much a swan can eat, can
prove rather expensive.

The key times when swans require more food than usual are
in early spring, just before breeding, when the female needs to
build up her fat reserves; during the post-breeding moult in late
summer, when the effort of replacing feathers also uses up energy; and, of course, during winter freeze-ups, when low tem-

peratures mean that lost energy needs to be replenished as rapidly as possible.

But for now, and for at least the next few months, there is plenty of food for both the adult swans and their cygnets, as they enter the season of plenty: summer.

Fig. 318 Der stumme oder Höcker-Schwan. (Olor mansuetus.)

3

SUMMER

London, thou art the flower of Cities all.
Above all rivers thy River hath renown,
Whose beryl streams, pleasant and clear,
Under thy lusty walls raineth down,
Where many a swan doth swim with wings fair . . .

William Dunbar, 'To the City of London' (1501)

London would not be the global metropolis it is today without the River Thames. The Romans, who founded the city soon after their successful conquest of Britain in AD 43, chose this site because the river – which they named *Tamesis* – gave them quick and easy access to the sea.

It's also fair to say that the Thames would not be the Thames without mute swans. From Cricklade in the west to the river's estuary roughly 350 km (220 miles) to the east, swans are often the most obvious and visible bird, gathering in loose flocks or pairs along the whole course of the river.

Mute swans have been present on the Thames since early medieval times, and probably much earlier. The office of the Keeper of the King's Swans (originally the Master of the King's Game of Swans, or Royal Swanherd) was established in the Royal Household some time during the thirteenth century. It continues to this day: though in 1993 the role was split into two, the Marker of the Swans and the Her Majesty's Swan Warden. Supported by three swanherds, Master's duties included conducting the annual count of the monarch's swans – known as 'swan-upping' – along the non-tidal part of the Thames each summer.

As long ago as 1496, the secretary to the Venetian Ambassador was writing in praise of the river's swans: 'It is a truly beautiful thing to behold one or two thousand tame swans upon the river Thames as I, and also your magnificence, have seen, which are eaten by the English like ducks and geese.'

Other foreign dignitaries were equally effusive at the sight of so many swans gathered in one place. One Italian visitor wrote in 1552 of the river abounding in flocks of swans, 'the sight of which, and their noise, are very agreeable to the fleets that meet them in their course', while a later writer described the 'broad river of Thames, most charming, and quite full of swans white as the very snow'. These accounts suggest that, elsewhere in Europe, large flocks of swans were not so commonplace.

But over time, the numbers of swans recorded on the River Thames – especially in central London – began to decline, as a result of a rapid rise in river traffic during the Industrial Revolution, and the consequent increase in water pollution from the many factories, wharves and docks sited along the river's banks.

By the mid-nineteenth century, the Thames's swan population had been much reduced. In the late 1860s, John Timbs, a prolific author and avid chronicler of London's history and natural history, wrote that 'the swans in the Thames are far less numerous than they used to be', though he also noted that 'the custom of swan-upping (vulgarly called swan-hopping) . . . is still observed, and is commemorated with high civic festivities'.

It still is, or at least it was until the summer of 2020 when, for only the second time in living memory, the census was cancelled – due, of course, to the impact of the Covid-19 lockdown and the resulting social distancing rules.

For Christopher Perrins, who has been the official Swan Warden since the role was first created in 1993, the cancellation was – even though fully expected – a blow. Though, as he points out, in July 2012 the event was also cancelled, following unseasonal flooding which meant the Thames was flowing too fast, and the water levels were too high, for swan-upping to be safely carried out.

Perrins is well suited to his role: an avuncular and charming man, he is also one of the world's most respected academic ornithologists. Following a long and distinguished career, he retired from his full-time role at the university in the early 2000s. He is now an Emeritus Fellow at Wolfson College, Oxford, and also at the prestigious Edward Grey Institute of Field Ornithology. He became interested in swans when trying to solve a problem that arises when scientists are studying bird population dynamics. With most species – especially songbirds such as the great tit, which he and his colleagues were studying at a wood near Oxford – you simply cannot find and track most of the birds present in a study area. The situation is further confused by what he calls 'floaters – birds that wander around the place and don't settle in territories – so you can never get a handle on them'.

The scientists sat and pondered whether there might be a species for which the majority of breeding birds could be tracked down more easily. 'We were trying to come up with a bird where we ought to be able to find all of them,' says Perrins, 'and guess what – a metre-long white thing fits the bill!'

The choice of the mute swan wasn't perfect – it turned out that some individuals are more elusive than you might think – but it was nevertheless a far better choice than any other species.

So, from the early 1960s onwards, Perrins and his colleagues began to put rings on mute swans in order to monitor their numbers – a timely intervention, given that at the time the species was undergoing a rapid decline, from a combination of poisoning by lead fishing weights and a series of very cold winters.

Thirty years later, with his lifetime's knowledge about the mute swan, Perrins was the natural choice for the appointment of Swan Warden. It is, like so many such posts, an unpaid role; as he ruefully remarks, 'I don't even get a swan at Christmas!'

*

Normally, swan-upping takes place over five days during the third week of July, between Sunbury-on-Thames in the east and Abingdon in the west, where the majority of the river's swans live. People from three ancient institutions take part: the Crown, under the supervision of Christopher Perrins and the Marker of the Swans, David Barber, and two 'livery companies' of the City of London, the Worshipful Company of Vintners (founded to control the selling of wine) and the Worshipful Company of Dyers (which controlled the sale of cloth). Both were granted the right to own swans sometime in the fifteenth or sixteenth centuries, and retain it to this day. The Crown team wear royal scarlet uniforms, the Dyers and Vintners, blue and white.

Although often portrayed as a quaint traditional ceremony, swan-upping is scientifically valuable, as it allows the ups and downs of the population, and the swans' breeding success, to be monitored. It is also very hard work. Perrins's role is as a kind of 'Swanfinder General', going ahead of the rest of the team in his motorboat to track down each pair and their offspring.

Each time a family is spotted, the traditional cry of 'All up!' is sounded, and the rest of the flotilla approach the birds, surrounding them to prevent their escape. The swans are not simply counted: their cygnets are also caught and marked – which, given their parents' size, and tendency to respond aggressively to any threat to their offspring's safety, is never an easy task.

Once each brood of cygnets has been caught, they are divided up, and then weighed, measured and ringed. The Crown (in the person of the Queen, who holds the ancient title 'Seigneur of the Swans') is deemed to own half of them, while the remainder are divided equally between the two livery companies, which add their own ring to each cygnet's leg to indicate who owns which birds.

Until the mid-1990s, the swans were also marked on their bill with a series of 'nicks', using a sharp knife to cut a specific pattern using a combination different of shapes, symbols and letters to ensure the swan could be matched with each owner; historically some owners marked the feet. Perrins recalls that when he started in his role as Warden the swans' bills were still being nicked – a surprisingly gory process:

> When the nicks were made on the side of the beak, it bled quite profusely, so they used to dab the bill with tar, which eventually stopped the bleeding. But by the time you'd got a cygnet covered with blood and tar, it wasn't very good PR! So, they were progressively making smaller and smaller marks, and after I'd been in the job for three or four years I managed to prove to them that the nicks they were making were so small that they were no longer visible when the birds became adults, and so they stopped doing it.

In *The Mute Swan in England*, N. F. Ticehurst features well over a hundred distinctive marks, while there may have been as many as 900 different ones in all. His diagrams reveal a wonderfully imaginative combination of designs, from simple triangular incisions to patterns showing the sun, stars, trees, keys, swords, ladders, a variety of crosses and a whole range of abstract designs. The oldest recorded mark dates back to 1370 – that of Sir Richard de Totesham of Kent, who kept his swans on Romney

Marsh – and some marks were in use continuously for several hundred years after they first appeared.

The custom also explains the rather unusual name of several pubs called 'The Swan with Two Necks', in locations including Bristol, Newcastle-under-Lyme in Staffordshire and Clitheroe in Lancashire. The name is, of course, not a reference to the star of some avian freak show, but a corruption of the word 'nicks'.

Surprisingly, perhaps, the Queen did not personally attend a swan-upping ceremony until 2009, almost six decades after she ascended the throne. She did so from the comfort of the oldest steamboat on the river, starting at Boveney Lock near Eton, and ending several hours later a few miles downriver, at Oakley Court near Windsor. In keeping with royal etiquette, Perrins is discreet about her response, though he does let slip that 'she seemed genuinely interested' in the event.

It is perhaps a pity that she did not attend earlier in her reign: in 1956, over 1,300 swans were recorded, but since then numbers have fallen dramatically. This is partly because of increased disturbance from people and boats on the Thames, but also down to the huge increase in gravel pits close to the river, which provide a quieter refuge for swans to raise their family. In recent years, the Thames swan population has begun to bounce back – partly, as Perrins points out, because of the ban in lead weights for fishing, which came into force in 1987. As a result, swan numbers rose dramatically in the next decade.

*

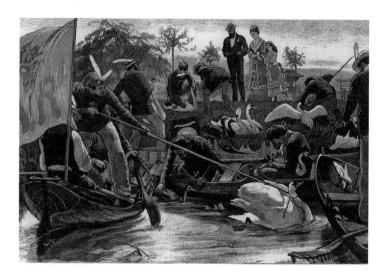

The differences between the enjoyable royal spectacle and the original purpose of swan-upping become clear when we dig a little deeper into its history. When the office of Master of the Swans was first created, back in the thirteenth century, it was a prestigious post, but also a very onerous one. Being so tame, large and valuable, swans were easy to catch and kill, so the Master – and his increasingly large team of deputies – needed to be on top of their job.

Swan-stealing was not the only problem: there were also frequent disputes between rival owners, which the Master and his team had to resolve. Certain rules were widely accepted: in the case where a male from one owner had paired with a female from another, the cygnets were divided equally; though if there were an odd number the extra one went – as we might expect from a patriarchal society – to the owner of the male.

With this position of high responsibility came great rewards. As well as his annual salary, the Master of the Swans was able to keep the monies raised by fines or penalties levied on miscreants and lawbreakers; not just those who killed swans or took their eggs, but also the revenue from illegal hunting or fishing within his domain. As Ticehurst comments, this 'must have provided a considerable incentive to carrying out their duties efficiently, and so promoted the preservation of the birds'. So not only did the Master and his deputies all benefit, but the swans did too.

Occasional disputes over the ownership of swans came to a head. In 1592, more than a century after the Act for Swans had been passed, two Dorset landowners, Dame Joan Young and Thomas Saunger, contested the legal right of the monarch – Queen Elizabeth I – to round up 400 unmarked swans on the Fleet (which they owned), and take possession of them. They claimed that they were the true owners of the birds, having earlier been granted the right to do so.

The legal arguments were complex and finely balanced, but the final verdict was clear: because swans were considered *ferae naturae* – wild animals – then Young and Saunger could not claim ownership of them. Therefore, the Crown did indeed, through royal privilege, have the rights over all unmarked swans on English waters. The Case of Swans (1592), as it is known, became a landmark decision in English property law.

Sadly, because in 2020 swan-upping was cancelled, due to the coronavirus lockdown, I was unable to witness the event for myself. This was not the first time my good intentions had been thwarted: back in the 1970s, I went down to the River Thames in my home village of Shepperton, on the day swan-upping was taking place. All I actually saw was a group of slightly drunken men emerge from the local hostelry, board their boats and head off into the distance.

I was delighted, therefore, to read Helen Macdonald's gripping account of the ceremony in her fine book of nature essays, *Vesper Flights*, published in 2020. Helen's interest had been kindled by two things: the Brexit vote of June 2016, and the depiction of swan-upping in a large and striking oil painting by the renowned English artist Stanley Spencer: *Swan Upping at Cookham* (Spencer's home village).

Now on display in London's Tate Britain gallery, this portrays a dramatic scene, with three adult swans trussed up with rope before they are marked. The story of the artwork is all the more poignant when we learn that Spencer started work on the painting in 1915, but did not finish it until after the First World War, in which he bravely served, had come to an end.

Unlike me, Helen did manage to witness the ceremony for herself, in July 2016. Her initial cynicism about what might appear to be a slice of the 'Merrie England' heritage industry, and her misgivings about a rise in racist violence since the Brexit vote, gradually changed to what she describes as 'a luxuriant, drunken joy':

Swan-upping is a progress in the old-fashioned sense, a journey upriver that claims the right not only to own swans but to own their meanings, the meanings of the river, the meanings of Englishness. You move through a landscape thick with narratives handed to you by others, and what you read from the banks as you pass is part of what you choose to believe about your nation and who you are.

In a wider sense, how we often view mute swans – as a symbol of Englishness – also straddles that tricky border between acceptable patriotism and ugly nationalism. That ambiguity is there in the way we regard swans as somehow 'different' from all other birds – being neither entirely wild, nor entirely tame – and also the way they are simultaneously celebrated for their beauty, feared for their violence and on occasion inexplicably killed.

Maybe this ambiguity is what makes the swan such an apt national symbol, in the light of the continued arguments about what it means to be English (as opposed to British) in a post-Brexit world.

The reverence and respect in which most people hold mute swans here in Britain is – despite the small minority who kill them – more or less taken for granted. So, it may come as a surprise to discover that, on the other side of the Atlantic, they are often regarded very differently: as alien invaders. Even there, though, they have their defenders.

In March 2015, officials from the New York State Department of Environmental Conservation announced a plan to cull more than 1,000 mute swans, by destroying nests and eggs and shooting the birds, in order to rid the state of what it described as a 'non-native and invasive species'.

Predictably, news of the slaughter outraged tabloid journalists in the UK: the *Daily Mirror* described the officials as 'public-funded animal assassins' and their plans as 'swan euthanasia'. In the USA itself, there was an immediate protest against the plans: a petition against them gained almost 10,000 signatures, while a local wildlife charity also condemned the cull as cruel and unnecessary.

This wasn't the first time that mute swans had been culled in the US. In November 2009, the animal welfare charity Mute Swan Advocacy condemned what it called the 'mass slaughter' of fifty-five swans at Mannington Meadows, a wetland reserve near Salem in New Jersey. The cull was carried out by the US Fish and Wildlife Service, the national agency that manages wildlife refuges, who justified it as part of a plan to prevent the potential spread of avian flu.

According to the world expert on non-native species, Sir Christopher Lever, mute swans were first introduced into North America – to the state of New York – in the late nineteenth century, with further introductions of more than 500 birds from 1910 to 1912. At first, they remained in a semi-captive state, but soon afterwards a number of individuals escaped and, slowly but surely, began to expand their range.

By the 1950s, mute swans could be found in Ohio, New Jersey and Rhode Island, and during the second half of the twentieth century they spread along much of the eastern half of the USA, from Minnesota in the north to Alabama in the south; while some birds also crossed the border to Canada.

Like many non-native species – and in a mirror image of Canada geese in Britain – mute swans have thrived in their new home, leading to trouble. From 1971 to 2000, the population on the Great Lakes doubled every seven or eight years, causing major environmental damage through over-grazing of aquatic vegetation such as eelgrass. They can also be aggressive to the two native swan species that overwinter in the USA: trumpeter and whistling swans.

As a result, in 2003 the US Fish and Wildlife Service made plans to reduce the mute swan population along the eastern seaboard by two-thirds, taking numbers back to the same level as the mid-1980s. The proposal won broad support from a wide range of wildlife and conservation organisations, including the various state Audubon Societies, but many animal welfare charities and individuals were passionately opposed to any action. Some even argued that the decision, in 2005, to remove federal protection from mute swans was wrong, claiming that the species was actually native to North America – a claim swiftly debunked by ornithologists.

One state where the species still enjoys protection is Connecticut. Nevertheless, in 2007 the Connecticut Audubon Society (the state's main bird conservation organisation) requested that the state Department of Environmental Protection remove mute swans from habitats along the coast, to prevent serious harm to the environment, and a continued reduction in biodiversity.

Predictably, perhaps, this riled the opposing side: Kathryn Burton, the founder of Save Our Swans USA, promised to fight a 'full-fledged war' against anyone who harmed the local swans. Save Our Swans does have a track record in taking legal action against other states who have tried to cull swans. But as the Audubon Society's director of science and conservation Milan Bull pointed out, 'Mute swans may be beautiful, but the havoc they wreak is anything but. They create a marine desert below the waterline and drive away native species.'

The state of Michigan has also had major problems with mute swans, which were first seen there in 1919, and have since grown to number 15,000 individuals – the most in any US state. Research has shown that in some wetlands, mute swans have wiped out the submerged aquatic vegetation completely, having a negative impact on the ecology and biodiversity of the lakes, and also harming the local economy: commercial and sports fishing are important industries in the area.

Further south, the city of Lakeland, on the tourist trail between Tampa and Orlando in central Florida, has also had a problem with its growing population of swans. But it has chosen a less drastic – and more creative – solution than culling.

Mute swans first appeared on Lake Morton, in the centre of Lakeland, in the 1920s, but by the early 1950s they had died out. Following news of their demise, a Lakeland woman living in England with her husband (who was serving with the US Air Force) wrote to Queen Elizabeth II to ask if the Crown could donate two swans to her home city, to replace those that had disappeared. Her Majesty duly obliged, and in February 1957 a pair of swans arrived, and were given a home on the city lake.

Soon swans became Lakeland's unofficial mascot, their images reproduced widely on T-shirts, and used as the logo of several local businesses.

During the following decades, swan numbers rose rapidly, with the consequence that there was not enough natural food for them. As a result, the city authorities were spending $10,000 every year on supplementary feeding. There was also a problem with wandering swans, which were venturing away from the lake to find new places to nest, and then being struck by motor vehicles and killed.

Eventually, the Parks and Recreation Department decided something had to be done. In a plan that could have been a story-line in the eponymous TV comedy series, they announced that they would sell off thirty-six of the city's eighty or so swans to the public, via a raffle.

Despite the cost of $400 a ticket – which the department claimed was 'below market value' – they were overwhelmed with enquiries. Some of the lucky new owners already had ponds where the swans could live and breed. Others placed them in wedding venues – appropriately given the swan's connections with love and fidelity – while, less aptly, some swans went to funeral homes. Steve Platt, from the city's Parks and Recreation Department – a man known as 'the Swanfather' – was clearly sad to see them go, telling the local newspaper, 'It'll be hard to say goodbye.'

In a poignant acknowledgement of the importance of swans as a symbol of enduring love and fidelity, in November 2020 a pair

of mute swans – named Chuck and Margie – were released at Longboat Key, a small town on an island just off the west coast of Florida. A ceremony was held to mark the release, which commemorated the lives and work of two local philanthropists, Charles and Margery Barancik, who had been killed in a car crash near their winter home on the island almost a year earlier.

4

AUTUMN

Swans always look as though they'd just been reading their own fan-mail.

Jan Struther, *Mrs Miniver* (1939)

The pair of mute swans, with their six well-grown cygnets, could be on any village pond, anywhere in the country. But they are, even by swan standards, major celebrities. Countless appearances on the BBC news, in national newspapers and on social media, a peak-time TV programme presented by the late Terry Wogan, and tens of thousands of visitors each year, all bear witness to their enduring fame. For these are the swans that ring a bell when they want to be fed.

On a late autumn afternoon, I had come to England's smallest city – Wells, in the heart of Somerset – to visit the thirteenth-century Bishop's Palace. I was greeted by Moira Anderson: officially, the Palace Administrator; unofficially, as she tells me with an engaging laugh, 'the Swan Whisperer'.

Brought up just down the road in Burnham-on-Sea, Moira confesses to having being rather frightened of swans before she took the job here back in 2011: 'I'd never been a huge fan of swans, as they are very big and frankly terrifying.' But once she started in her new role, and got to know the resident pair of swans, Moira's fear soon turned to enthusiasm for these magnificent birds. She discovered that not only did the adult swans, Brynn and Wynn, regularly ring the Palace bell to get food themselves, but year after year had trained their cygnets (a total of forty-five in all) to. Like baby blue tits watching their parents pecking open the foil tops of milk bottles to get the cream, it seems that young swans are quick learners.

In early 2019, for her annual appraisal, Moira's official objectives read as follows:

(i) Get a new pair of swans for the moat.
(ii) Teach them to ring the bell for food.
(iii) Make sure they have fluffy cygnets.

The tone of urgency referred to events during the previous spring of 2018. Just after the pair had mated, Moira noticed that Brynn, the male, was looking decidedly unwell. He was caught and taken to a local animal rescue centre, Secret World, but sadly died overnight. At the time, his mate Wynn had just laid five eggs; incredibly, she continued to incubate them until they hatched, and went on to successfully rear all five cygnets to adulthood.

That autumn, things took another unexpected turn. By then, four of the five cygnets had already left; soon afterwards, Wynn and the remaining cygnet departed as well. This was the first time Wynn had vanished in the whole time Moira had worked at the Palace, and the first time she had ever seen her fly.

Wynn did return briefly once more, the following January, for three days. Moira believes that she came back to check whether Brynn was there; once she realised he was not, she then left again. 'I've since discovered that what we should have done when Brynn died was to bring his body back for Wynn to see him for the last time, to get some kind of closure.'

Moira was advised to wait three months to ensure that Wynn really had gone for good. Once that time was up, she contacted swan rescue centres to find out if any of them had a suitable pair that could be released onto the Palace moat. Finally, in May 2019, the new male and female were delivered, courtesy of Swan Rescue South Wales. 'They arrived with great fanfare, because we'd not had any swans here for six months, and everyone in Wells wanted to see them back where they belong.'

The Palace asked people to suggest names for the new birds, and had over 300 suggestions – some of which, as Moira admits, were not terribly suitable (previous submissions had inevitably included 'Swanny McSwanface'). Eventually, having got together over coffee and cakes, the Palace staff chose Gabriel for the male and Grace for the female – each name appropriate to this Christian setting.

Now, though, Moira and her colleagues faced a race against time. In previous years, the Palace swans had usually mated in March, with the female laying her clutch of eggs soon afterwards, and the cygnets hatching in early May. When this new pair finally arrived, they were at least two months behind schedule, and it looked as if they would not manage to raise a brood at all that summer.

But to Moira's delight, they mated immediately, after which Gabriel built a nest and Grace laid four eggs. Sadly, only one egg actually hatched, but the lone chick – aptly named Lucky – did manage to survive and fledge to adulthood.

Meanwhile, just a few days after they had arrived, Gabriel had already learned to ring the bell for food, much to everyone's relief, though Grace has still not done so. Moira believes that may be because, some time before she arrived at the Bishop's Palace, Grace and her original partner became entangled with a rope, which led to his death, making her understandably cautious around humans. Grace does, however, readily take food, which is provided in the form of special 'swan pellets', far more nutritious than the usual white bread.

Moira is keen to stress that the swans remain wild, with their wings unclipped, so they are able to leave at any time. But as she points out, 'This place is like a five-star hotel – very safe, with food always on tap – why on earth would they ever fly away?'

In 2020, things went more or less back to normal. In early March, Gabriel built a nest on a small island in the middle of the moat, right next to a bridge:

This pair are not great at deciding where to nest – the place they chose was right next to what is usually a very busy footpath! One of the few benefits of lockdown was, of course, that there were no visitors, so she was able to sit quietly, without hundreds of people peering down at her every day. I am hoping that next year, though, they find a more secluded place to nest.

In 2020, Grace laid a higher-than-average clutch of eight eggs, all of which hatched, though one cygnet emerged several days after the others, by which time it was alone in the nest. As Moira explained, the next few hours were very traumatic for her and her colleagues, who were working from home because of lockdown, and had to witness the drama unfolding via a remote nest camera. 'We watched this tiny, new-born cygnet as it struggled to survive, while its parents and siblings were out of sight on the Palace moat. Sadly, there was nothing we could do, and it died later that day.'

Even then, the drama was far from over. One of the remaining seven cygnets decided to go for an adventure, and went over the nearby sluice, ending up two miles downstream, where a local woman noticed him 'having an argument with a magpie' close to a new housing estate. Aptly named Gulliver, he was checked over by a local vet and brought back to the moat the next day.

Unfortunately, as often happens when a baby bird has been separated from its parents, Gabriel and Grace rejected him, and so he had to be sent to a rescue centre for his own safety.

LE CYGNE.

Moira offered to take me to meet the swan family, and witness Gabriel ringing the bell for food. But swans are just as unpredictable as any other wild birds, and when we emerged from the Palace gatehouse, where the bell is attached, they were of course nowhere to be found. As we turned the corner of the moat, however, I caught sight of them in the distance, surrounded by a noisy flock of black-headed gulls and a smaller crowd of human admirers.

As we approached, I could immediately see that the six cygnets were already well advanced in losing their dark juvenile plumage. Though their heads and necks were still dusky grey, and the primary wing feathers buffish-brown, their backs, flanks and tail were now almost completely white, with just a few darker patches remaining.

I suspect that the cygnets' partial adoption of their white plumage, which takes place from autumn into winter, acts as a signal for the parent swans to kick them out of their territory. As Moira explained, in October the adults begin to get aggressive towards their offspring, and for the next few months, into the New Year, they become more and more antagonistic, until the last of the brood (usually the females) finally depart in February or March. Even as we watched them peacefully feeding, against the distant backdrop of the medieval cathedral, one cygnet got a bit too close to its parent, who immediately shooed it away.

'The best thing about the moat is that it is so long and straight,' Moira explained, 'which means that when the youngsters are learning to fly they have a really long run-up, pattering their feet along the surface until they finally get airborne – an incredible sound.' This made me reflect on how on earth my local cygnets ever manage to fly, given the lack of suitable watery runways on the moor behind my home.

I casually mentioned to Moira that, unlike the subjects of my previous bird biographies – the robin, wren and swallow – swans don't really do very much for most of the time, apart from loafing around. Her answer made me realise I had got this completely wrong:

You just have to be patient. Over the years, almost everything I've learned about swans has been through spending time with them, observing what they do. For instance, the female will teach the young cygnets to feed, by stirring up the gunk from the bottom of the moat. And when, on occasion, another pair of swans has arrived here, it's like a Mexican stand-off on the lawn! Each pair is different, too: Wynn and Brynn always stayed put, whereas Grace and Gabriel often wander around Market Square, to the delight of local residents.

Her account of the wandering swans sounded somehow familiar. Then I remembered why. In the 2007 British film comedy *Hot Fuzz*, two hapless policemen, played by Simon Pegg and Nick Frost, take part in a memorable scene in which they try to catch a rogue swan that has gone missing from 'the castle'. The film was shot in director Edgar Wright's home town of Wells and, for the scene featuring the uncooperative swan, right here in the grounds of the Bishop's Palace.

I said my goodbyes to Moira, promising to come back to see Gabriel ringing the bell sometime soon. As I walked past the fifteenth-century Swan Hotel – its name a timely reminder of the central place of these birds in our culture – I wondered how, and when, the custom of training the Bishop's Palace swans to ring the bell originated.

Fortunately, one of the volunteers at the Palace, Rosemary Cooke, has looked into the historical origins of the bell-ringing swans. Her investigations suggest that the custom began in the middle of the nineteenth century, at a time when the incumbent Bishop of Bath and Wells was Lord Auckland – or, to give him his full name and title, Robert Eden, 3rd Baron Auckland. He was bishop from 1854 until he resigned, aged seventy, in 1869, a year before his death.

Lord Auckland had a large family, neatly divided into five sons and five daughters. Evidence points strongly to the fact that one of the girls – either Eleanor, Emily, Emma, Florence or Maria – was the first to have taught the swans to ring the bell. Rosemary bases this on her discovery of a poem by William Catcott, the self-styled 'Baker Bard of Wells', a popular poet who was living in the city at the time.

In 1858, a swan on the Palace moat had been killed, leading to the following appeal in the local newspaper, the *Wells Journal*:

> Last week some evil-disposed person or persons cruelly beat one of the swans on the moat and the bird has since died.
>
> What could have actuated the perpetration of such a wicked act? It is too difficult to conceive.
>
> A reward of 2 guineas is offered by the Lord Bishop, the owner of the birds, for the apprehension and conviction of the offender and we sincerely trust the scoundrel or scoundrels will not long evade detection as this is not the first act of cruelty that has been committed upon the swans on the moat which surrounds the palace.

This appeal inspired William Catcott to write his poem, with the splendidly melodramatic title 'Lines on a Beautiful Swan Barbarously Beaten to Death on the Palace Moat':

No more upon the water clear
Wilt thou sweet bird, thy form display,
Nor to the slippery margin steer,
To take the offered crumb away;
Nor e'er again, with instinct rare,
Thy daily wants and wishes tell,
Making the wondering gazers stare,
By striking out thy dinner bell.

No more beneath the shady trees,
Wilt thou unfold thy snowy wing,
And glide along with graceful ease,
To beautify the spot, poor thing!
No more around the turrets grey,
Wilt thou thy morning's circuit take,
Nor fold thy wing at close of day,
Upon the bosom of the lake.

The last line of the first stanza – 'By striking out the dinner bell' – clearly shows that the swans had by then already been trained to ring the bell for food. Supporting evidence comes from a note about Lord Auckland at the Wells and Mendip Museum, on nearby Cathedral Green, which reads: 'His daughter [sadly, which one is not specified] is remembered locally as the lady who taught the swan to ring for their food . . .'

Later on, several references suggest that one of the six daughters of the next incumbent of the bishopric, Lord Hervey (bishop from 1869 to 1874), might also have taught the swans. However,

Rosemary believes the confusion may have arisen because a new pair of swans needed to be shown how to ring the bell.

Whoever first trained the swans at the Bishop's Palace to do so, it is rather delightful that today – almost two centuries later – the custom continues.

Brynn, the previous male swan at the Bishop's Palace, was twelve or thirteen years old when he died; a fairly typical lifespan for a mute swan. Swans have just a 40 per cent chance of surviving their first year, and a two in three chance of making it through their second year. Once they have reached maturity, though, they have a roughly 80–90 per cent chance of surviving each year thereafter.

A typical lifespan is ten years, but one bird ringed as a cygnet at Abbotsbury in Dorset on 12 September 1980 was caught again just 2 kilometres away on 23 October 2009, having reached the age of twenty-nine years, one month and eleven days. This is very impressive, but does not come close to the estimate provided by Thomas Bewick in 1804: 'It is the generally received opinion that the Swan lives to a very great age, some say a century, and others have protracted their lives to three hundred years!'

A sixteenth-century contemporary of William Caxton, the printer and publisher Wynkyn der Worde, went even further. Using some rather dodgy maths, he calculated that because a man lives for eighty-one years, a goose three times as long as a man, and a swan three times as long as goose, then swans must therefore live for 729 years. He may have been good at arithmetic, but logic was clearly not his strong point.

So how do swans die? Well, apart from the constant threat of deliberate and accidental killing by human beings and their dogs (see Chapter 2), they can meet their fate in a number of ways. These include starvation during harsh winters, pollution by chemicals and oil discharged into waterways, flying into large obstacles, and another factor I have already touched on in relation to the recovery of swans in the River Thames: lead poisoning.

Towards the end of the 1970s, residents of and visitors to Stratford-on-Avon – home of William Shakespeare, 'the Swan of Avon' – became concerned at the virtual disappearance of mute swans from the town. A working group, including Christopher Perrins, reported that the loss was almost entirely down to the continued use of lead weights by fishermen.

Writing in 1990, Janet Kear takes up the story: 'Anglers were outraged; they pointed out that they had been using lead weights for well over 100 years on waters traditionally inhabited by mute swans, and that their relationship was a happy one.'

The anglers claimed that the swan population crash on the River Avon was instead down to a gradual deterioration in their riverine habitat, caused by a combination of loss of waterweed on which the swans feed, an increase in boat traffic, and the trend for river 'improvement' schemes, which often achieve the very opposite of what has been intended, at least for wildlife. The fishermen even suggested that the custom of local people and tourists feeding swans on stale bread might be what was killing them.

But the science was indisputable: roughly four out of every ten swans subjected to a post-mortem had died of lead poisoning. It was estimated that nationally, between 3,370 and 4,190 swans might be dying every year – a substantial proportion of the total UK number. Despite these horrific figures, it took almost a decade before lead weights were finally banned, and even then, anglers were allowed to continue to use those already in their possession.

Yet one puzzle remained. Why, if local anglers had been using lead weights for more than a century, had the swan population only recently begun to show serious decline?

The answer lay in the type of line they were using. Before the invention of nylon, just before the Second World War, thrifty anglers would always have taken their cotton or linen lines, and the weights attached to them, home after fishing. But in the post-war era, nylon fishing line became so cheap that if a line became snagged, the angler simply cut it off and left it where it was – with the lethal hooks and toxic lead weights attached.

Staggering amounts of lines and weights are routinely discarded: almost 4,000 pieces of lead, and almost 5 miles of nylon line, have been picked up along a single short stretch of river. Fishing line, too, is lethal to swans, as shown in November 2020, when a young swan was rescued after becoming entangled in a fishing line – complete with weights – at a nature reserve in Fife. This is just one of many similar cases regularly reported in the media.

The good news is that in Stratford-on-Avon itself, the swan population is now thriving. In the summer of 2020, the annual count led by the town's mayor (and lifelong swan enthusiast) Cyril Bennis totalled a new record of 150 swans, the most since he first began the survey in 1981.

Incidentally, as a result of his work proving the danger from lead weights, Christopher Perrins was later appointed a Lieutenant of the Victorian Order for his services to swans (an honour he modestly failed to mention when I interviewed him for this book).

Fishing weights are not the only issue when it comes to this toxic heavy metal. Lead pellets can still be legally used to shoot gamebirds – including some species of wildfowl – which means that many wetlands are contaminated with these lethal objects, despite the fact that non-toxic alternatives have long been available.

Again, the figures are astonishing: in a typical year, shooters fire more than 6,000 tonnes of lead ammunition. Given that a typical shotgun cartridge contains roughly 200 separate pieces of shot, this means there are more than thirty *billion* individual fragments of lead discharged into the countryside annually – almost 500 for every man, woman and child in the UK.

The Wildfowl and Wetlands Trust, which has long campaigned for a ban on lead shot, estimates that as many as 100,000 ducks, geese, swans and other waterbirds are being killed every year by ingesting lead shot, which they accidentally eat when picking up grit to aid their digestion.

That figure equates to a mortality of roughly one in ten individuals across all species; but for whooper and Bewick's swans the figure is much higher, with roughly one in four dying from consuming lead shot. Across Europe as a whole, the WWT estimate that one million waterbirds are dying each year, with a further three million suffering ill health. With populations of swans already under threat from other factors, this is likely to have a seriously negative effect on long-term numbers.

The WWT cleverly used social media to spearhead its campaign: asking people on Twitter and Facebook to post images of swans to persuade the members of the European Parliament to endorse the ban on lead shot. Alongside stunning photographs of male and female swans making the classic heart-shaped pose, these also included a striking image of psychedelic swans, evoking the work of the pop-art guru Andy Warhol.

It worked. In autumn 2020, European Union lawmakers finally took the long-awaited decision to ban lead shotgun pellets from being used in and around wetlands. Dr Ruth Cromie, a Research Fellow at the WWT, welcomed the decision, though she rightly sounded a note of caution:

> It remains to be seen how the UK government will respond, but this is the beginning of the end of lead ammunition and the start of a healthier, greener future for Europe's wildlife and people. After decades of pollution and suffering, lead ammunition may finally be consigned to history.

In the light of Brexit, the oft-parroted slogan of 'taking back control', and the current government's bias towards landowning and shooting interests, I am not holding my breath that lead shotgun ammunition will be banned in the UK any time soon.

*

In 1990, Janet Kear presciently suggested that 'Now that the purchase of large lead fishing weights is effectively banned, it may be that electricity-carrying cables strung across their flight path will be the swan's greatest threat.' She pointed to the fact that in the Hebrides, where lead poisoning is unusual because of the relatively low number of anglers and shooters, more than four out of five swan deaths (where the cause was known) were a result of collisions with electricity wires, the birds being either fatally injured or electrocuted. The good news is that measures can be taken to reduce the deaths from such collisions.

In the springs of 2004 and 2006, surveys revealed that a total of thirty mute swans had been killed after colliding with power lines at Abberton Reservoir in Essex. During the summer of 2006, to reduce this heavy death toll, 500 bright red 'flight diverters' – each measuring 32 by 17.5 cm (12.5 by 6.9 inches) – were placed 5 metres apart along a 1.5 km stretch of lines. The following spring, just one swan was killed, and in 2008 none died. Since then, similar diverters have been used at sites all over Britain, with a massive reduction in swan deaths.

Were Janet Kear writing now, though, she might be shocked at the number of collisions between swans (and other low-flying birds) and another series of large structures erected by humans: wind turbines. These can be huge: the largest in the UK, at Lowestoft in Suffolk, reaches a total height (from the base to the tip of the blade) of 126 metres (413 feet). That's considerably taller than London's St Paul's Cathedral.

The turbines' design and function, with the blades sweeping round at up to 192 mph (120 mph), also make it hard to avoid bird strikes. Offshore wind farms, in particular, could pose a threat to our two species of migratory swans, whooper and Bewick's, especially as they travel in family parties with inexperienced youngsters making their first journey.

One of the reasons swans are especially vulnerable to such collisions is that, with their eyes on either side of their heads, they have a 'blind spot' immediately in front of them. This may also explain why, in January 2021, an elderly woman in Nottinghamshire had a lucky escape when a fully grown mute swan crashed through her bathroom window just moments after she had left the room. Fortunately, the swan escaped with a wound to its wing, and was stitched up and taken to an animal rescue centre to recover from its adventure.

Being so large, swans are not killed by predators as often as their smaller relatives. Nevertheless, they can fall victim to foxes. However, as Perrins and Birkhead point out in their monograph *The Mute Swan*, 'it seems . . . unlikely that a fox could catch a healthy swan, since the birds are normally on open water or in the middle of open fields'.

They go on to suggest that the vast majority of swans reported as having been eaten by foxes were either scavenged, having died or been killed by other means, or were already sick or injured when the fox came across them. In some cases, the swan may have fallen victim to one of the many avian diseases and parasites to which they can be susceptible, weakening their defences against this opportunistic predator.

*

What effects this catalogue of man-made and natural hazards has on the overall population of the mute swan, especially in the long term, is hard to calculate. According to David Cabot, whose New Naturalist volume *Wildfowl* is a fascinating compendium of facts, figures and scientific analysis, the rapid increase in numbers during the 1980s and 1990s was probably down to a reduction in mortality among both young and adult swans.

Cabot suggests that mild winters – and therefore more widely available food – might have led to an increased survival rate among juvenile swans, while this, along with a reduction in lead poisoning and possibly more food provided by humans, would have reduced the death toll for adult birds too.

Because swans are long-lived, even if they do suffer a fall in numbers due, say, to an unexpectedly harsh winter, they are still able to recover during the following few years. This may also, he suggests, be because when adult mortality does rise, there are always young non-breeding birds 'waiting in the wings', ready to take their place as adult breeders.

Although we usually regard the mute swan as a very sedentary species, Cabot notes that in autumn and winter they do move short distances away from where they were hatched and raised. Typically, though, when young birds leave their parents they travel an average of just 14–16 kilometres (roughly 9 or 10 miles) from where they were born, though a small number – just over two per cent – venture more than 80 km (50 miles) away.

In the northern parts of the mute swan's range, however, the species sometimes travels much further: in Russia and China, the breeding populations of mute swans must migrate each

autumn, to avoid winter temperatures that drop well below freezing for months on end, making it much harder to find food.

Prolonged periods of harsh winter weather may also force swans in northern Europe to head south and west. William Yarrell writes of the events of January 1838, when, during severe conditions, flocks of swans were observed flying southwards along the east coast, from Scotland to the mouth of the Thames. And during the winter of 1962/63 – the infamous 'Big Freeze' – several mute swans ringed in Scandinavia and the Low Countries were later found in south-east England, having fled the snow and ice for marginally milder climes. Some survived, and later returned to their breeding grounds.

★

In his popular collection of avian folklore *Bird Facts and Fallacies*, published in the mid-1920s, Lewis R. W. Loyd makes a confident assertion about the cultural importance of swans: 'It is probable that in folklore and mythology, particularly those of Europe, the Swan has figured more largely than any bird that flies.'

He may well be right. The skylark and the nightingale loom large in both Greek and Roman texts and English poetry, and when it comes to British folklore the robin and the wren are not far behind. Yet all three European species of swans are virtually ever present in mythologies and literature, in many ancient and modern cultures all the way across northern Europe, Asia and North America.

As far back as the earliest-known civilisations, and through to the present day, swans can be found in cave paintings and fairy tales, pagan rituals and religious iconography, music and litera-ture, astronomy and advertising, pop songs and pub signs, and countless other aspects of both our cultural and day-to-day lives.

Swans have also been associated with a number of very differ-ent characters: from the Swan Maiden of the nomadic peoples of Siberia (and many other northern cultures), through the Swan Knight of King Arthur, to the central character of the ballet *Swan Lake*; few if any other species of bird have such a wide range of human personifications.

Lewis Loyd examines one of the most persistent of the many myths surrounding mute swans: that, after a life spent without uttering vocal sounds, they sing before they die. This fallacy goes all the way back to at least the sixth century BC, when it appears in Aesop's fable 'The Swan and the Goose'. The next mention comes more than a hundred years later, 458 BC, in Aeschylus's play *Agamemnon*. In a dramatic scene, the eponymous king's wife Clytemnestra – who has just murdered both her husband and his mistress Cassandra – describes the latter as a swan who 'hath sung her last lament in death'.

In his pioneering *History of Animals*, written in the fourth century BC, Aristotle writes that swans 'are musical, and sing chiefly at the approach of death', an assertion repeated by the Roman poet Ovid. But another Roman, the philosopher Pliny the Elder, writing in AD 77, was not convinced, stating that 'observation shows that the story that the dying swan sings is false'. Another sceptic was the poet Samuel Taylor Coleridge, who wrote this brief and waspish epigram, 'To a Volunteer Singer':

> Swans sing before they die; 'twere no bad thing,
> Should certain persons die before they sing.

The conceit of the swan singing as it dies may relate to the long-held notion that these pure white birds are a living embodiment of the souls of the dead: a concept known as 'metempsychosis'. The author and polymath Sir Thomas Browne, writing in 1646, quoted Plato on this theory, described as 'the transmigration of the souls of men into the bodies of beasts most suitable unto their human condition'. Thus, after his death, the musician and poet Orpheus became a swan.

The image of the dying swan singing as it expires appears throughout English literature, too, as in these lines from one of Geoffrey Chaucer's minor poems:

But as the Swan, I have herd seyd ful yore,
Ageyns his dethe shall singen his penance.

Shakespeare, too, includes it in his long poem 'The Rape of Lucrece':

And now this pale Swan in her watery nest,
Begins the sad dirge of her certain ending . . .

And in *The Merchant of Venice*, Shakespeare's noble heroine Portia utters these powerful lines about her suitor Bassanio:

Let music sound while he doth make his choice.
Then if he lose he makes a swanlike end,
Fading in music.

The myth of the swan bursting into song at the moment of death was bound to appeal to the Romantic poets, and so it did. Here is part of Alfred, Lord Tennyson's poem 'The Dying Swan':

The wild swan's death-hymn took the soul
Of that waste place with joy
Hidden in sorrow: at first to the ear
The warble was low, and full and clear . . .
But anon her awful jubilant voice,
With a music strange and manifold,
Flow'd forth on a carol free and bold;

As when a mighty people rejoice

With shawms, and with cymbals, and harps of gold . . .

Tennyson refers to the 'wild swan', perhaps implying that the species involved is not the mute swan at all, but the whooper, whose loud flight call has been variously described as 'honking', 'bugling', 'clanging', 'trumpeting' and of course 'whooping'.

The eighteenth-century Prussian naturalist Peter Simon Pallas did point out that when a whooper swan dies, its looped trachea can cause it to utter a long, drawn-out note; a phenomenon also observed in a dying tundra swan, which, unlike the mute swan, has a similar loop in its windpipe. So maybe in this respect the myth is true, after all.

Given that Pyotr Ilyich Tchaikovsky, the composer of the celebrated ballet *Swan Lake*, was Russian, I have always thought that the species at the centre of the story might actually be a whooper swan. After all, the whooper's breeding range extends all the way across northern Russia, where the mute swan is not found, so it would perhaps have been a more natural choice.

Swan Lake may be the best-known musical composition inspired by swans, but the Finnish composer Jean Sibelius had an even more immediate moment of inspiration, again featuring whooper swans.

His epiphany occurred on the morning of 21 April 1915, when he looked out of the window of his home, north of Helsinki. At that precise moment, on the lake in front of his house, he saw sixteen whooper swans taking flight. He described this special moment in his diary: 'Today at ten to eleven I saw 16 swans. One of my greatest experiences! Lord God, what beauty! They circled over me for a long time.'

Sibelius used the sight and sound of the swans to create what has been described as 'the greatest finale of all time'; to end his 5th Symphony. The final movement opens with fast semiquavers on the violins, before a majestic French horn melody glides in. It was this theme that Sibelius called his 'swan hymn', and it is an echo of both the graceful movements and haunting calls of the whooper swans that flew over his home, so evocative of the Finnish landscape of lakes and forests.

Gayle Wood, a nature writer based in Cornwall, describes her response to one of her favourite pieces of classical music:

> It is an exhilarating musical depiction not only of their sound but of their flight, of those huge birds taking wing, beating the air. There is a sense of the wild in the music that speaks to a sense of the wild in all of us.

This was not the first time Sibelius had been inspired by the sound of whooper swans: twenty years previously, in 1895, he had composed a 'tone poem' called 'The Swan of Tuonela', based on a nineteenth-century Finnish epic poem; he also wrote the music to accompany the Swedish playwright August Strindberg's work *Swanwhite: A Fairy Drama*.

*

LACERTA. CYGNUS. LYRA. VULPECULA AND ANSER.

Look far up into the night sky, and there is even a swan there. All the way across the northern hemisphere, especially in summer and early autumn, the constellation Cygnus is easily visible along the Milky Way, roughly 1,470 light years (almost 9,000 *trillion* miles) from Earth. Despite its enormous distance from our own galaxy, Cygnus is easy to see on a clear night, because its brightest star, Deneb or Alpha Cygni (the nineteenth brightest star in the night sky), is somewhere between 50,000 and 200,000 times brighter than our own Sun. *Cygnus* is the Latin word for swan, and the constellation was so named because the most prominent stars form the shape of a cross which, with a little imaginative leeway, could be said to represent a long-necked bird, such as a swan, in flight.

The name derives from the Greek myths: Cygnus, the son of the sea god Neptune, is supposed to have grieved so much over the death of his friend Phaeton that the sun god Apollo transformed him into a swan, and sent him into the heavens. This in turn relates to the myth that Apollo would travel to the land of the north wind each spring, in a chariot pulled by flying swans.

In yet another connection, Zeus – who had killed Phaeton for riding his own chariot too close to Earth – is said to have observed Leda (Queen of Sparta and mother of Helen of Troy) bathing, and turned himself into a swan in order to approach and then seduce her. The resulting myth, Leda and the Swan, is one of the most enduring and also most controversial of all ancient stories.

Through the centuries the story has featured in artworks, songs and poems by – among others – Leonardo da Vinci, Mi-

chelangelo, Cézanne, Lou Reed and W. B. Yeats, whose sonnet 'Leda and the Swan' is charged with erotic imagery:

> A sudden blow: the great wings beating still
> Above the staggering girl, her thighs caressed
> By the dark webs, her nape caught in his bill,
> He holds her helpless breast upon his breast.
>
> How can those terrified vague fingers push
> The feathered glory from her loosening thighs?

The power of the myth is thought partly to derive from its ability to depict the longstanding taboo of sexual intercourse in visual art and poetry, by substituting the swan for the man. The swan may have been chosen because its long, sinuous neck could be seen as a phallic symbol.

In recent years, however, the myth of Leda and the Swan has been examined and questioned in both poetry and literary criticism from a feminist perspective, which argues that male artists and critics have romanticised the act of rape at the heart of the myth, and that Leda is degraded, rather than empowered, by being violated.

This argument is summarised in a 2017 paper 'Re-tellings of the Myth of Leda and the Swan: A Feminist Perspective', by Shadi S. Neimneh, Nisreen M. Sawwa and Marwan M. Obeidat:

> It is suggested, therefore, that these detrimental effects of rape on
> Leda and her deteriorating psychological state are due to being

violated by a male deity, and thus Leda has fallen victim to rape as well as patriarchy.

The more we look into the myths and legends surrounding swans – and the works of art and literature that perpetuate them – the more we become aware of our essential ambivalence towards these birds. We regard them simultaneously as calm and beautiful, yet also aggressive and often violent; we put them at the centre of a tragic ballet, yet also in a poem about rape; and while some people love them with a passion, others cruelly kill or injure them.

Swans continue to inspire composers, artists and writers. The news that a swan on London's Hampstead Heath had finally found a mate (named Wallace) four years after being widowed, led to the swan rescue volunteer Louisa Green and illustrator Dee McLean to create a children's book, *There's Something About Wallace*, published in 2020. Subtitled 'Lockdown love on Hampstead Heath', this tells the tale of how, long after her first mate died when he crashed into a nearby building, Mrs Newbie (as she had been christened by locals) pairs up with a swan at a rescue sanctuary, before being returned to the heath where she lays seven eggs. Even then, the drama isn't quite over: having been attacked by a dog, she is taken into care once again, leaving her mate to raise the cygnets on his own.

On the other side of the Atlantic, Gregory Maguire, author of the bestselling *Wicked* series of books (now turned into an equally successful musical), has recently published a new fantasy novel for young adults: *A Wild Winter Swan*. Inspired by another swan-based Hans Christian Andersen tale, about a young girl whose eleven brothers are turned into swans by their wicked stepmother, it updates the story to modern-day America. Reviews have been largely positive, with one describing the book as 'lush prose, laced with humour and poignancy, weaving the fabulous into the quotidian world'. It seems that at dark times like these we all need fairy tales.

Even darker times – the horrors of the First World War – inspired another powerful swan-based work of art: a song written in Scottish Gaelic. 'The White Swan' (*An Eala Bhàn*) was written by the soldier-poet Donald MacDonald, from the Hebridean island of North Uist.

Born in 1887, MacDonald – also known by the name Red Donald of Corunna (*Dòmhnall Ruadh Chorùna*) – began writing poetry in the year 1900, aged just thirteen. At seventeen, he enlisted in the militia, and a decade later, in 1914, joined the Cameron Highlanders and went off to fight on the Western Front.

While serving there, he wrote a number of poems in his native tongue, many expressing his grief and sorrow for fallen comrades, and at the horrors of the war. But his best-known composition, 'The White Swan', was the result of his own close brush with death. In autumn 1916, fighting in the Battle of the Somme, MacDonald came under artillery fire. A shell exploded close to where he was hiding, and knocked him unconscious. Having regained consciousness he tried to escape, but was wounded by a bullet from a German sniper. Finally, several hours later, he slowly crawled back to safety.

Eventually, he was sent back home, where he wrote the song that would lead to his lifelong fame. 'The White Swan' is a lament for his sweetheart Maggie MacLeod and his home on the island – made all the more poignant by the fact that, although he did eventually return to North Uist, he and Maggie never married.

Composed in the oral tradition, the song and his poems were not recorded and published until much later in his life, when Fred Macaulay, a BBC producer also from North Uist, persuaded MacDonald to read and record them just before his death, in 1967.

'The White Swan' is still performed by well-known Scottish singers and bands including the folk group Capercaillie and the singer Julie Fowlis, whose numinous performance is guaranteed to make the hairs on the back of your neck stand on end. The lyrics remain moving to this day, more than a century after it was first written:

> From the very time that I left
> The high bens of the mist
> The little glens of dalliance
> Of the lochans, the bays and the forelands
> And the white swan dwelling there
> Whom I daily pursue.
>
> Maggie, don't be sad,
> Love, if I should die
> Who among men endures eternally?

> We are all only on a journey . . .
> I am not fated to survive.

Yet survive he did. His song remains as a tribute to the many who did not.

Nowadays the word 'swansong' (or 'swan song') is often used metaphorically to describe a final performance, not only by a singer but also in many other fields from sport to literature. This usage appears to have come into the English language from the German compound noun *Schwanengesang*, the title of a collection of songs written by the Austrian composer Franz Schubert towards the end of his life, and published in 1829, a year after his death. This it likely to be an echo of that longstanding myth that swans sing at the point of death.

More recent examples of 'swansongs' include David Bowie's final album *Blackstar*, also released following his death in 2016, and described by many reviewers as 'the perfect swansong'. The term 'swan song' features too in the title of songs by artists Lana del Rey and Dua Lipa, released in 2015 and 2019 respectively. Swan Song Records was also the name of the record label launched in 1974 by the rock band Led Zeppelin, its logo, an image of a human figure with a swan's wings, based on a painting by the nineteenth-century American artist William Rimmer.

Another Seventies rock icon, Marc Bolan, penned 'Ride a White Swan', which became his first hit (for his band T. Rex, in 1970) and, according to one critic, inadvertently kick-started glam rock. When Bolan died in a car crash seven years later, the most prominent floral tribute at his funeral was a massive white swan – symbolising, perhaps, his perceived fragility and beauty. Even today, more than forty years later, devoted fans still leave model swans at the roadside shrine in Barnes in south-west London that has grown up at the site of the crash.

The concept of the 'swansong' features regularly on the back pages of newspapers, to mark the final game by a much-loved sportsman or woman, especially if they have ended their career in triumph by winning a tournament. These include the late US basketball player Kobe Bryant, who scored 60 points in his final game for the Los Angeles Lakers; the footballer Xavi Hernandez, ending his long and illustrious career at Barcelona by lifting the European Champions League trophy; and the cricketer (and later Pakistan's Prime Minister) Imran Khan, who in 1992, in his very last match as captain of his country, led his team to defeat England to win the World Cup.

Less widely known myths about swans discussed by Lewis Loyd include the belief that swans' eggs hatch out during thunderstorms; the idea of 'swan-maidens' (prominent in folklore in countries as far apart as Iceland and Australia); and the widespread use of the swan in heraldry and place names: 'itself sufficient indication of the regard and even reverence in which the swan was held'.

In Scotland, Loyd notes, swans are 'looked upon as good prognosticators of the weather', with the early arrival of whooper swans on Orkney indicating a cold winter – a belief confirmed by the Victorian folklorist Richard Inwards, whose 1898 volume

Weather Lore is a veritable treasure trove of meteorological sayings and proverbs.

Another Victorian folklorist, the Revd Charles Swainson, writing in 1885, refers to the belief that breeding swans can forecast heavy rainfalls: 'the birds whose home is on the banks of the Thames raise their nests so as to save their eggs from being chilled by the water'.

Swans also feature prominently in ancient mythologies, cultures and religions all around the northern hemisphere. A story from Norse mythology seeks to explain the swan's coloration, which supposedly came about when a pair of swans drank from a well whose water was so pure and holy that any creature that touched it immediately turned white. This has a legacy today in the phrase to symbolise mutual co-operation within the 'Nordic Council', the nations of Norway, Sweden, Finland, Denmark and Iceland, which are collectively known as the 'Five Swans'. The Council's flag features a stylised depiction of a white swan, on a blue disc, against a white background.

Swans are also central to Hinduism, in which they are regarded as saintly, as although their feathers come into contact with water they do not get wet. The swan is linked to the Hindu goddess Saraswati (also known as Sharada), who represents a wealth of virtues including wisdom, knowledge, art, speech and learning, and is usually depicted dressed in pure white, with a swan at her feet, symbolising spiritual perfection. Because of this, Saraswati is sometimes referred to as Hamsavahini, a name deriving from the word *hamsa*, meaning swan.

*

In popular culture, the frequent depiction of swans in advertising and marketing must be down to a combination of their striking appearance, white colour and especially their association with purity, as we have already seen with the KLM campaign.

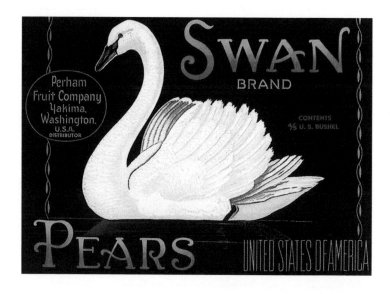

In 1941, Lever Brothers launched a revolutionary new soap with the brand name 'Swan', which they claimed was so pure it could be used on the sensitive skin of infants and babies. In the USA, where sponsorship of radio broadcasts was permitted, Swan Soap funded several popular entertainment shows, including those presented by the comedians George Burns, Gracie Allen and Bob Hope. The company also invested in artists to produce images for their advertising, all of which prominently featured a swan, which later became collectors' items.

Other products and services using images of swans include knitting patterns, brooms, flour, travelcards (to allow the user to 'swan around' London) and, most famously, matches. Swan Vestas matches first appeared (under the brand name 'Swan') in 1883, made and distributed by the Collard and Kendall match company in Bootle, Lancashire (now Merseyside).

In 1906, when the company merged with Bryant and May, the matches were renamed 'Swan Vestas', presumably after Vesta, the Roman goddess of fire and the hearth. During the inter-war years, they were marketed as 'Britain's best-selling match', and up to at least the 1970s as 'the smoker's match' – until the link between smoking and lung cancer made that particular slogan rather less attractive.

One day, sometime during the 1930s, a young bird photographer named Eric Hosking received a telephone call from a London printing firm, Sir Joseph Causton and Sons. The request was straightforward: could he send them a photograph of a swan, for an unnamed client? Hosking tells the story in his 1970 memoir, *An Eye for a Bird*:

> I promised to put it in the post that day, but on looking through my files was dismayed to find that none was suitable. Somehow or other the promise must be kept, so, armed with bread crusts and camera, I cycled to the pond at Waterlow Park near Highgate Hill where there were several swans.

Struggling awkwardly to throw the bread towards the birds with one hand, while trying to fire his camera shutter with the other, Hosking finally succeeded in taking a number of portraits of a fine adult mute swan. He then cycled back home as quickly as he could, developed the pictures and made some prints, which he sent off that very evening.

His speed and determination paid off, as he later recalled:

> One of these pictures appeared shortly after on hoardings all over the country. A drawing made from my photograph has been the Swan Vesta sign ever since.

I have already noted (in Chapter 3) that several pubs in England go by the name 'the Swan with Two Necks' – a corrupted form of 'two nicks', from the marks made on the bird's bill during swan-upping. There are also many pubs and hotels named 'the White Swan', and 'the Black Swan', including a Michelin-starred 'restaurant with rooms' in the village of Oldstead, a few miles north of York, which offers a ten-course tasting menu for a mere £125.

But of all the many pubs named after a species of bird, by far the commonest is simply the Swan, with more than 400 licensed premises. Depending on which figures are used, this comes fourth or fifth in the league table of pub names, after the Red Lion, Crown, Royal Oak and White Hart. Like these, the names derive from the signs first erected outside pubs during the reign of Edward III in the mid-fourteenth century, to allow their customers – the majority of whom would have been unable to read – to identify their chosen hostelry through a visual symbol. The Swan may have been chosen because it is one of the easiest of all birds to portray and identify.

Janet Kear notes that in some cities there were so many hostelries that qualifying names were added to distinguish one 'Swan' from another – in York, as well as the Swan, there were also pubs named the White Swan, the Old White Swan, the Cygnet and, several centuries before this species was actually discovered, the Black Swan (also known as the Mucky Duck).

We often think of avian folklore as something from the distant past, created by ancient cultures and then handed down over many generations to the present day. But in the distant future, cultural historians and archaeologists may look back on the twenty-first century as a golden age of swan folklore, judging by the many examples – some in the classic tradition, others new and surprising – I have come across while researching this book.

Swans are often used as mascots – notably for the only Football League club nicknamed after the species, Swansea City, whose fluffy match-day mascot is called Cyril the Swan. Cyril has at times brought positive publicity to the club, as when he was voted 'Best Mascot' by readers of *BBC Match of the Day* magazine. He also enjoyed a romantic interlude when he 'married' Cybil the Swan at their old ground in 2005, to mark the move to a new, purpose-built stadium. But at times, the nine-foot-tall Cyril has attracted less welcome publicity, when his enthusiasm for the club has spilled over into violence.

Famously, during a match against Millwall in 2001, Cyril had an altercation with the London club's mascot Zampa the Lion. During a disputed penalty, Cyril (or more accurately, Swansea's groundsman Eddie Donne) pulled off Zampa's head and kicked it across the pitch, taunting his rival with the words 'Don't fuck with the Swans!' As a result, the club was fined £1,000, and a code of conduct for mascots was drawn up.

With their huge size and unmistakable silhouette, swans also lend themselves to the visual arts. In August 2020, twenty-five enormous wooden swans began to appear around the Staffordshire town of Burton-on-Trent. The swans, which had been painted in a variety of colours and patterns by a group of new and emerging artists, were part of an arts trail sponsored by local businesses, intended to promote the town to locals and visitors One of the artists, illustrator Heather Horsley, went the extra mile – or seven – by doing a sponsored run around the trail, along with members of her running club.

Afterwards, the swans were auctioned off to raise more than £12,000 for the local YMCA, whose income had been hugely reduced because of the Covid-19 lockdown, to fund their work with people sleeping rough in Burton.

In the Australian city of Perth, famous for its black swans, an even bigger swan – 7 metres long and 5 metres high – has been created out of 10,000 plastic bottles, to promote a major plastics recycling campaign. Plastic drinks containers make up almost half of all litter across Western Australia, so the 'cash for containers' scheme, in which each bottle carries a refundable deposit of 10 cents, is timely and much needed.

At the other end of the financial scale, plans have also been announced to design and build a new super-luxury megayacht in the shape of a swan, complete with 'feathers' and a long, curved neck. Created by the Lazzarini Design Studio in Rome, the *Avanguardia* (from the Italian for avant-garde) was inspired by a 1970s Japanese anime series, in which a robot swan pretends to be a woman. The shapely, movable prow at the front of the 137-metre craft is not just for show: it also houses the yacht's control tower. The price? A cool $500 million (approximately £375 million).

5

WINTER

The Icelanders, whose year may be said to consist but of one long day of summer months, when they enjoy the light of the sun, and one long night of Winter, when he never cheers them with his rays, compare this cry of the wild swan to the sound of a violin; and when heard at the end of their long and dreary Winter, announcing the approach of genial weather, it is associated and coupled in their minds with all that is cheerful and delightful.

Edward Stanley,
A Familiar History of Birds (1865)

Towards dusk on 9 December 1967, a radar controller in Northern Ireland noticed an unusual 'echo' over the Inner Hebrides, just off the west coast of Scotland. The signal was moving very rapidly south, at a speed of about 75 knots (83 miles per hour). Knowing that there were no meteorological balloons in the area, he wondered if this had come from a flock of migrating birds, but the altitude of between 8,000 and 8,500 metres (roughly 26,000 to 28,000 feet) was surely far too high.

Curious, he asked the pilot of a civilian transport aircraft to take a look. The pilot came back to confirm that they were indeed birds: a flock of about thirty swans. The radar controller continued to track the swans until they descended towards Lough Foyle, in Northern Ireland. This was – and remains – the highest altitude at which any birds of any species have ever been recorded in UK airspace.

Given the time of year, the flight path and the destination, it has long been assumed that this was a late-arriving flock of whooper swans, heading south from their Icelandic breeding grounds to spend the winter in the comparatively balmy climate of the British Isles. However, given that whooper swans general-ly migrate at very low levels, and only occasionally rise to heights of up to 1,700 metres (5,900 feet), the idea that they might reach an altitude five times greater does seem unlikely.

The late Colin Pennycuick, the world expert on the mechanics of bird flight, was intrigued by this apparently bizarre sighting, and investigated if and how it might have occurred. He was un-able to draw a definitive conclusion, but he did observe that the swans might possibly have taken advantage of an unusual atmos-pheric phenomenon known as 'lee waves', which can provide enough lift to allow birds to soar to great heights. That appears to be the only possible explanation for the 1967 observation.

*

Though a few pairs of whooper swans do breed in Britain and Ireland – mainly in the far north of Scotland and north-west Ireland – this is insignificant compared to the numbers arriving here in late autumn and early winter: an estimated 34,000 birds. That's close to half the winter population of mute swans, and dwarfs that of the whooper's smaller relative, Bewick's swan, of which fewer than 5,000 overwinter here.

Whoopers – named, of course, after their loud, whooping call – are, along with Bewick's, known as 'wild swans', in contrast to the 'tame swan' tag once given to our resident species. The whooper was also the first swan species to be officially described, by the Swedish taxonomist Linnaeus, in 1758 – hence its scientific name: *Cygnus cygnus*.

Both whooper and Bewick's swans have a yellow-and-black (rather than orange-and-black) bill: in the case of the whooper, the yellow extends below the nostrils; for the Bewick's, it ends above.

Whoopers are about the same size as the mute swan: 152 cm (60 inches) long, with a slightly longer wingspan of 2.3 metres (roughly 7.5 feet), reflecting the longer journeys they need to undertake. They are also generally lighter, weighing an average of 9.3 kg (20.5 lb), though this does vary considerably, as William Yarrell notes. He writes that 'Colonel Hawker' (Peter Hawker, a noted wildfowler) frequently shot whooper swans along the Hampshire and Dorset coasts between Lymington and Poole Harbour:

> When struck in the body, they become helpless from their weight
> . . . No birds vary more in weight than Hoopers [sic]. In the last
> winter, 1838, I killed them from thirteen to twenty-one pounds.
> On one occasion I knocked down eight at a shot . . . and they
> averaged nineteen pounds [8.6 kg] each.

Whooper swans have an enormous breeding range, the largest by far of any of the world's swan species. This extends across the northern taiga (or boreal) zone for almost 30 million square kilometres (over 11 million square miles) – equivalent to the land areas of the USA, Canada and China combined – though the birds do not breed throughout the whole of this area. Its sibling species, the (black-billed) trumpeter swan, has a much smaller breeding range, being mostly confined to western North America. The world population of whoopers is currently estimated at 248,000 individuals.

Whooper swans breed from May through to August, from Iceland in the west to the Kamchatka Peninsula in the east, and as far south as central Asia. Like mute swans, they lay a single clutch, usually four or five eggs, which are incubated by the female for up to six weeks; the young fledge between eleven and fourteen weeks later – just in time to head south and west for the winter.

In the northernmost parts of their range, way beyond the Arctic Circle, whoopers face an annual race against time to complete their 130-day breeding cycle; if the snow and ice melt late in the spring, they may be unsuccessful. Somewhere between 35 and 40 per cent of whooper pairs attempt to breed each year in the milder climate of Iceland, but this falls to just 5 per cent in the far north of Finland, where the whooper swan is the national bird, featuring on the country's one-euro coin.

Virtually all whooper swans are migratory, wintering across much of temperate Europe and Asia, from Ireland to Japan, and as far south as the eastern Mediterranean and Caspian seas. The birds that overwinter in Britain and Ireland almost all come from Iceland (though a small number of Icelandic whoopers do stay put for the whole year).

To get here, these birds must fly as far as 1,200 km (750 miles) – the longest sea crossing undertaken by any swan species. They usually travel in family groups, to help the youngsters find their way south.

This sea crossing sometimes ends in failure, especially if they face bad weather or strong southerly winds. However, the birds can 'ride out' tricky conditions by landing on the surface of the sea; hence the crossing can take anything between eleven hours and four days.

The wildfowl expert David Cabot welcomes the annual arrival of these 'magnificent and magical' birds near his home in Ireland:

When a fresh party of swans arrives on a lake, already occupied by other whoopers, there follows a most boisterous and frenzied display of wing flapping and head pumping, accompanied by a cacophony of wild callings, while the swans face each other as if in combat.

But, as he goes on to point out, this is mostly for show – what he calls 'a ritualised greeting or triumph ceremony' – almost as if they are celebrating their collective arrival.

During the winter, whoopers usually gather in large flocks, often in the company of both mute and Bewick's swans. Key sites include Caerlaverock and the Solway Firth in the Scottish Borders, the Ribble Estuary in Lancashire, the Ouse Washes on the Cambridgeshire / Norfolk border, and various sites in Northern Ireland including Lough Foyle, Beg, Erne and Neagh.

Family parties can also be seen at other wetland sites across much of the UK (though with a distinct bias towards the north). They can overwinter on surprisingly small waterbodies: back in the 1980s, I would regularly see a flock of whoopers on a farm pond in the Forest of Dean. Like other swans, their diet is mostly vegetarian: they feed mainly on farmland, on improved pasture and arable crops, such as turnips, potatoes and winter wheat.

Numbers of whoopers wintering in Britain have risen dramatically since the turn of the century, with record counts at the Ouse Washes – the most important site – of more than 5,000 birds, reflecting a steady rise in the Icelandic breeding population. Back in the early 1960s, the total wintering number for the whole of Britain and Ireland was roughly the same number as can now be found at this one location.

Having arrived here in October or November, whoopers head back north in March or April, often having reinforced their pair bond by displaying enthusiastically, and very noisily, to one another before they leave. They bob up and down, calling loudly in pairs and larger groups, creating a memorable spectacle.

Whooper swans face similar threats to other large wildfowl: lead poisoning, collisions with power cables, pollution, habitat loss and deliberate killing. For centuries they were also hunted for food and feathers on their breeding and wintering grounds, especially when moulting – and therefore flightless – towards the end of the breeding season, when up to 300 birds a day might be collected.

Thomas Bewick, writing at the start of the nineteenth century, vividly describes the many uses to which the various peoples of North America and elsewhere put their quarry:

> The flesh is highly esteemed by them as a delicious food, as are also the eggs, which are gathered in the spring. The Icelanders, Kamtschatdales [inhabitants of the Kamchatka Peninsula in the Russian far east], and other natives of the northern world, dress their skins with the down on, sew them together, and make them into garments of various kinds . . . while the larger feathers are formed into caps and plumes to decorate the heads of their chiefs and warriors.

Eventually, whooper swans were virtually wiped out in parts of their range, including western Greenland and Japan, though numbers have since managed to recover thanks to conservation measures. In Japan, where the species is known as 'the Angel of Winter', they are now welcomed.

Whoopers continued to be hunted legally in Iceland until 1913, when they were given legal protection, which they also have in Britain. Nevertheless, illegal killing still occurs throughout their range: a study in which Icelandic swans wintering in the UK were given X-rays revealed that roughly one in seven had lead shotgun pellets embedded in their bodies.

In October 1916 the Irish poet W. B. Yeats wrote one of his best-known poems, 'The Wild Swans at Coole', soon after a visit to Coole Park, in County Galway. He compares his mood as he approaches old age (he was actually in his early fifties), with how he felt on an earlier visit, in 1897, as a younger man. Bewick's swans are very rare winter visitors to Ireland, and more or less unknown in the far west, so we can be sure that the birds Yeats is writing about are whoopers.

The poem begins with a simple but elegant description of the scene:

> The trees are in their autumn beauty,
> The woodland paths are dry,
> Under the October twilight the water
> Mirrors a still sky;
> Upon the brimming water among the stones
> Are nine-and-fifty swans.

But the serenity of the verse soon takes a darker turn, as the poet reflects on the passing of time and the coming of old age:

> I have looked upon those brilliant creatures,
> And now my heart is sore.
> All's changed since I, hearing at twilight,
> The first time on this shore,

The bell-beat of their wings above my head,
Trod with a lighter tread.

Then, at the close, he muses on his sense of loss when they depart north again:

Among what rushes will they build,
By what lake's edge or pool
Delight men's eyes when I awake some day
To find they have flown away?

The moral of the poem seems to be that while his own life is only going in one direction – towards old age and death – the swans' annual cycle of departure and return is eternal.

Yeats's poem is the best known in a very long tradition of poems featuring these impressive birds, the earliest being the Anglo-Saxon poem known as 'The Seafarer', written towards the end of the seventh century AD. The poem contains this compelling image of 'wild swans' (presumably whoopers) heading north in April:

There I heard nothing but the roar of the sea,
Of the ice-cold wave, and sometimes the song of the wild swan . . .

Those two poems, more than 1,200 years apart, reveal the staying power of all swans – not just the familiar mute – in our culture. In Norse mythologies, the myths about whooper swans go back even further, according to the ornithologist Mark Brazil, whose lifelong passion for this species shines through in his comprehensive 2003 monograph, *The Whooper Swan*.

Brazil reveals that when whoopers headed away from their breeding grounds in autumn, it was believed they were flying all the way to Valhalla, home to warriors slain in battle. When they returned the following spring, they were greeted as the symbol of the end of the long, dark winter and the coming of summer; a welcome they no doubt still receive in these high northern latitudes.

The whooper's smaller cousin, Bewick's swan, has been equally celebrated in more recent times – but as the signal of the coming of winter, not spring; and not in the far north, but right here in Britain, as the story of the twentieth century's greatest conservationist reveals.

★

Peter (later Sir Peter) Scott was one of those enormously talent-
ed and gifted men who seem to have stepped out of the pages of
a *Boy's Own* adventure story. In addition to an unparalleled career
in wildlife conservation, he was an accomplished artist, a deco-
rated wartime naval officer who designed a revolutionary new
way of camouflaging battleships, and an Olympic medallist – he
won a bronze for sailing at the 1936 summer games.

Son of the celebrated polar explorer Robert Falcon Scott –
'Scott of the Antarctic' – Peter never knew his father, who per-
ished just a few miles short of safety in March 1912, when his son
was just two years old. Famously, Captain Scott's last letter to his
wife, written when he knew he would never be returning home,
exhorted her to 'make the boy interested in natural history'. She
did, and Peter Scott went on to become arguably the most important and influential figure in the long history of our human
interactions with birds and wildlife.

Today, Scott's best-known legacy is the Wildfowl and Wetlands Trust, whose headquarters is at Slimbridge, on the Severn
Estuary in Gloucestershire. It was here that he, along with his
wife Philippa and daughter Dafila, began one of the longest-
running studies of any wild bird ever undertaken: that of the
Bewick's swan.

Bewick's swans have not always overwintered on the Severn
Estuary: until Scott began the Slimbridge project, just after the
Second World War, they were a rare visitor there. Then, during
the winter of 1948/49, two years after the centre opened, a
wild male Bewick's swan landed within the grounds, presumably lured in by the presence of a captive whistling swan in the

wildfowl collection. Scott managed to capture it, then acquired a mate from the Netherlands, and the pair – named Mr and Mrs Noah – settled down to breed.

During the next decade or so, wild Bewick's swans began to overwinter regularly on the nearby saltmarshes, alongside the River Severn. By the winter of 1963/64, about twenty were coming in pairs and family parties to visit the ponds in the enclosures at Slimbridge.

That winter, Scott had the idea of moving his small captive flock of Bewick's and whistling swans (now numbering seven birds) onto the lake in front of his house, in the hope of luring the wild Bewick's in, so that he could study and paint them. But he could hardly have imagined what would happen next. The birds were moved on 9 February 1964, and the very next day, the first wild Bewick's swan appeared. Dafila Scott takes up the story:

> The following day, more wild ones flew in and my father noticed that their bill patterns differed from the first. So, he painted detailed watercolours of the bill patterns of each swan, with views from the front and each side, and gave the individuals names. Thus began a remarkable study that continues to this day.

As long ago as 1923, the ornithologist C. N. Acland had noticed that the bill pattern of Bewick's swans varied, but the Scotts were the first to realise that these small but visible differences could be used to identify individual swans. They gave them names to facilitate recognition, and to study their behaviour, especially their relationships with one another, at a level of detail not normally possible with wild birds.

In 1964/65, a total of sixty-eight individuals were recorded at Slimbridge during the course of the winter, with a peak count of fifty-five birds on one day, and the Scott family recognised many of the swans that they had seen the previous winter. As the years went by, more and more swans were identified, painted, named and logged, until by 1989 – the year of Peter Scott's death – the database contained 6,700 individual birds seen at Slimbridge. That figure has since risen to over 12,000.

Thanks to the regular provision of food, and the safe environment of the lake, numbers continued to rise during the 1960s, then stabilised until the very cold winter of 1978/79, when a total of 610 Bewick's swans was recorded on a single day, and 721 different birds were counted during the winter as a whole. This reflected not only the severe weather conditions in continental Europe pushing the swans further west, but also an increase in the north-west European population at the time.

The Slimbridge study of Bewick's swans remains one of the longest-lasting and most comprehensive studies of any bird, anywhere in the world. It also gave rise to studies of the species elsewhere in its wintering range, such as the Netherlands and Germany.

For many years, relatively little was known about the swans' lives on their breeding grounds in the Russian Arctic but, with the cessation of the Cold War, expeditions to study Bewick's swans on their nesting territories at last became possible. In 1991 Dafila and the swan researcher Eileen Rees first visited the

Nenetskiy region of the Russian Arctic, a main breeding area for Bewick's swans wintering in Europe, and a long-term collaborative study with Russian colleagues ensued.

Yet without that first crucial observation, and realisation that each bill pattern was unique and individually identifiable, we might never have unravelled so many of the mysteries of the complex life cycle of this charismatic and beautiful bird.

On the shortest day of the year, I arrived at Slimbridge an hour or so before sunset. As I walked up to the entrance, it dawned on me that it was just over fifty years since I had first come here, with my late mother, on what was our first 'proper' birding trip outside the London suburbs. It was, I recall, the late October half-term, and we were either too early to see the wintering swans and geese, or simply didn't know where to go. Today, I hoped, I would have better fortune.

With half an hour or so before the start of the daily swan feed, I had time to pop into the centre's well-stocked shop. Among the books and gifts, I came across a range of products with the Scotts' famous design of the Bewick's swans' bills: mugs, notebooks, even fridge magnets, each showing off the informative series of yellow-and-black bill patterns. I later learned that the master bedroom in Peter Scott's home is now wallpapered with the swan bill design.

The very same pattern is also displayed on the wall of the Peng Observatory at Slimbridge, named after the Singapore conservationist Yuen-Peng McNeice, and opened by the Queen in 1988. The observatory looks out onto the Rushy Pen, where Peter Scott first noticed the swans' distinctive bill patterns. As if on cue, as I entered, a pair of Bewick's swans flew in to land right in front of me, and began displaying.

I watched with delight as they raised and lowered their heads and necks in synchrony, opening their wings and uttering their deep, haunting calls. Like many wildfowl, Bewick's stay together all year round, not just on their breeding territories; and this display is a way of both reinforcing their lifelong pair bond, and establishing the pecking order with other pairs, as these new arrivals then proceeded to do.

The daily swan feed, which on this, the day of the winter solstice, took place at 4 p.m. – more or less bang on sunset – always attracts hundreds of other birds. The most prominent, apart from the swans, were several hundred greylag and Canada geese: noisy, feral birds that stay here all year round.

Some of the smaller birds here – shelducks, tufted ducks, pochards, shovelers, mallards and pintails – are also local residents, but many others would have travelled some distance, from their breeding grounds in Scandinavia and eastern Europe. All were eagerly awaiting the signal that the feed was about to begin.

Beforehand, I chatted to Scott Petrek, reserve warden at Slimbridge. Scott is passionate about these beautiful birds, and showed me the three basic patterns, each named after the Old English word for 'bill'. The first and commonest of these, accounting for about 60 per cent of the birds, was 'yellow neb', in which the yellow extends right across the top of the bill; followed by 'dark neb', where the yellow is divided by a strip of black; and 'penny-face', in which there is a round yellow patch roughly the size of an old penny, bordered by black, in the centre of the bill. When the swans conveniently turned to face me, I soon managed to distinguish all three patterns.

Scott also pointed out three brownish-grey cygnets, out of six that had returned this year. One was with a single female, Chapillary, whose mate has presumably died; the other two were with both parents, Sarindi and Sarind. The female of this pair was very unusual in that she is one of only two Bewick's swans ever recorded here to have 'divorced' her partner, and re-paired with another male, while the original mate was still alive.

Both the number of swans overall (sixty so far this year) and the proportion of young to adults (roughly 10 per cent of the total) were well below what the WWT would normally expect. In recent years, only about 100 Bewick's have usually overwintered at Slimbridge, a number itself well down on what there used to be. This is because of a combination of a global population decline, and milder winters, which mean that many birds stay put further to the east.

The ratio of cygnets to adults needs to be above 18 per cent to maintain numbers long-term, so again, this is a cause for concern. Fortunately, these are long-lived birds – the record age is twenty-nine, and several birds here today were in their twenties – so there is still hope that their decline can be turned around.

In the past, the first Bewick's swans used to arrive at Slimbridge in the last week of October. This has now shifted back a few days to the first week of November, with new birds still turning up as late as Christmas week. If there is a cold spell on the near-continent, or in eastern England, more may arrive at any time during the winter.

One of the birds we were watching had a collar on, which sends location information in real time, enabling the WWT team to track the bird's movements. This can throw up some surprises. One young female, Maisie, left her roost in the Netherlands at 7 o'clock one morning, crossed the North Sea to arrive at the WWT centre at Welney in the Ouse Washes by 9 a.m. and, after a four-hour flight west, reached Slimbridge later that same afternoon.

Weather permitting, Bewick's swans usually begin to head back east and north from Valentine's Day (14 February), and the stragglers have mostly all left by the end of February. Again, with milder winters, they are tending to head off earlier than they used to, generally preferring to migrate on south-westerly tailwinds.

However, in February 2021 a group of twenty Bewick's swans had to abandon their eastward journey due to bad weather. The flock had set off in the second week of February, and the birds were not expected to return until the following autumn. But as they headed back towards their breeding grounds, they ran into Storm Darcy – dubbed the new 'Beast from the East' – a weather system bringing sub-zero temperatures and heavy falls of snow across eastern England. A few days later, a dozen swans returned, including one individual never before seen at Slimbridge – aptly named Darcy after the storm that had brought him.

On my December visit, once enough birds had gathered out-side the picture windows of the observatory, Scott finally headed outside to feed them. As he distributed shovelfuls of wheat, the geese and shelducks barged to the head of the queue, like eager buyers elbowing their way to the front at the winter sales. True to their more serene and reserved nature, the Bewick's swans stayed further back, waiting for Scott to make his circuit of the lake.

They can afford to be patient, as their long necks will enable them to reach the food that has fallen to the bottom; they are also able to use their feet as paddles to stir it up, before dipping their heads beneath the surface to feed. All the while, they were accompanied by the two species of diving duck – tufted and po-chard – which can also reach food in these slightly deeper waters.

The swans, geese and ducks are normally given a second feed at 6.30 p.m., before they go to roost on the water, where they are safe from predators such as foxes; and another at 8 o'clock the following morning, after which they head out onto the sur-rounding fields to graze for the rest of the day. They then return like clockwork just before 4 o'clock each afternoon.

Half an hour later, the feed was over. As I got up to leave, I looked back at the flock of fifty or more Bewick's swans, illumin-ated by the floodlights against the blue-grey waters of the lake. I fervently hoped that these birds do not disappear as a wintering bird here in Britain during my lifetime.

*

BEWICK'S SWAN
Cygnus bewicki, *Yarr.*

Ironically, one well-known bird book that makes no mention at all of the Bewick's swan is Thomas Bewick's *A History of British Birds*. In the second volume, 'Water Birds', published in 1804, he devotes several pages to 'the Wild Swan', but it is clear that he is referring to what he describes as 'the Elk, Hooper, or Whistling Swan' – i.e. the Whooper.

This is not as odd as it sounds. Bewick's swan was not actually named until 1830, two years after the death of the eponymous ornithologist, engraver and political radical. The species had only been discovered the year before, when William Yarrell examined the remains of a small swan, and realised that it was sufficiently different from the familiar whooper to be pronounced a new species, which he formally named after Thomas Bewick. The original specimen remains on display today in the Hancock Museum, in Bewick's home city of Newcastle upon Tyne.

Bewick's swan was, as Barbara and Richard Mearns note in their 1988 book *Biographies for Birdwatchers*, 'almost the last large bird occurring in Britain to be recognised and described by the scientific community'. That is presumably because of its superficial resemblance to its larger cousin, even though the fossil record reveals that Bewick's swans have been present in Britain since the Middle Pleistocene epoch, roughly half a million years ago.

Bewick's swans are considerably smaller than either mute or whooper swans. Adults are on average 121 cm (roughly 48 inches) long, with a wingspan of 196 cm (77 inches). As in the other swan species, males are larger and heavier than females, weighing 6.4 kg (14 lb) and 5.3 kg (11.7 lb) respectively – just over half the weight of their larger relatives. They are shorter-necked and stockier than the whooper swan, with the same straight neck. While the bill is also yellow and black, the black extends considerably further up the bill.

The 'superspecies' to which Bewick's belongs, the tundra swan, is the world's most northerly breeding swan species; virtually all nest north of the Arctic Circle. The North American subspecies *columbianus*, which was discovered in 1806 by pioneering explorers Meriwether Lewis and William Clark, is known as the 'whistling swan', from its high-pitched call.

The whistling swan – whose bill is almost completely black, with what has been described as a 'yellow tear streak' running below the eye – breeds in northern Canada, Alaska, the Aleutian Islands and the extreme east of Siberia, where its range overlaps with the slightly smaller Eurasian subspecies *bewickii*: our familiar Bewick's swan. In the past, Bewick's and whistling swans have been separated into two full species, based on their small differences in appearance. However, as they occasionally interbreed in the wild, most scientists now consider them to be distinct subspecies rather than separate.

Bewick's swan breeds across a broad swathe of northern Siberia west to the Kola Peninsula, in the far north-west of Arctic Russia, near the borders with Norway and Finland. As with the whoopers, the timing of the breeding season is crucial; perhaps, given the more northerly latitudes at which it breeds, even more so. The female lays a single clutch of three to five eggs, which she incubates for roughly thirty days; the cygnets fledge between forty and forty-five days later.

Breeding success is highly dependent on the spring weather conditions from year to year, and the speed at which the thaw comes to their breeding grounds on the Arctic tundra. In poor years, more than two-thirds of adults will not breed at all, instead spending the summer in large flocks to keep safe from predators.

Adult and young Bewick's swans, along with their eggs, may be taken by Arctic foxes, snowy owls, rough-legged buzzards, white-tailed eagles, gulls and skuas, while in Alaska, brown bears predate on whistling swans. As we have already seen with other swan species, they may also fall victim to collisions with power lines or wind turbines, lead poisoning, and especially illegal hunting. Scott Petrek told me that some wildfowl hunters in Russia deliberately target Bewick's swans, because they wrongly blame them for recent falls in duck numbers.

In North America, each autumn, whistling swans head south

to spend the winter along the eastern and western coasts of the United States, while Bewick's swans leave their breeding grounds in late September and early October, heading either south-east, south or south-west, depending on where they breed.

Eastern birds (which make up the bulk of the global population) overwinter in China, Korea and Japan, while western birds come to Europe, notably the Baltic, the Low Countries and the UK, with some heading directly south to the Caspian Sea. They travel in 'extended family' parties, including not only that year's young, but sometimes also those from the previous year or two.

Despite breeding further north than whoopers, Bewick's swans in Britain tend to winter in the south: mostly in England, at sites such as the Ouse and Nene Washes, and the Severn Estuary – including, of course, Slimbridge. They also overwinter in Belgium, France, Denmark, and increasingly in Germany.

However, in recent years, numbers have fallen considerably: the north-west European population numbered 30,000 during the 1990s, but fell to 18,000 by 2010; meanwhile, numbers wintering in Britain and Ireland have dropped from 7,200 in 2000 to just 4,400 in 2015.

This is partly due to milder winters to the east, allowing the birds to stay put on the other side of the North Sea; over the past fifty years the winter range has shifted eastwards by an average of 13 km (roughly 8 miles) a year. But the fall in numbers is not just down to changes in the climate; it also reflects a serious and worrying decline in the birds' breeding success.

The number of Bewick's swans wintering in Britain continues to vary from winter to winter, depending on the prevailing weather conditions elsewhere in northern Europe, but the long-term trend is clearly down. But now, the plight of the Bewick's swan has been highlighted by the efforts of a remarkable woman: the conservationist and campaigner Sacha Dench.

In autumn 2016, after two years' careful planning, Sacha Dench set off on an extraordinary, and potentially life-threatening, adventure. She took to the air in a motorised paraglider, at the start of a journey from Arctic Russia to Britain, a distance of 7,200 km (4,500 miles), which would take her through the airspace of eleven different countries. Her aim was to travel alongside the Bewick's swans' migration route, as they made this same hazardous annual journey.

As Sacha explains, there was no time to lose:

> We needed a way to really quickly understand exactly what was going wrong in different places on the birds' migration, try to fix it, and get a lot more people outside conservation to help. What better way to do this than fly with them and see the threats from their point of view, which would also attract the public attention we needed. This was without doubt a pretty unusual approach, but the swan decline was happening so fast that WWT and others along the flyway were prepared to back it.

She hoped to meet people who encounter the swans along their route, to try to discover why the birds are in decline, rally support from people to help them, and draw attention to their plight via the world's media. At the start of her journey, she would often land among nomadic communities in northern Russia, who would share their food, and their stories about the swans, with her. And in each country she visited, local conservationists used her appearance to highlight the plight of Bewick's swans.

What made her achievement all the more extraordinary was that she had originally learned to fly a paraglider to get over her fear of flying, following a terrifying experience in a small aircraft in Panama. Although she never had to use her parachute, she did have to make frequent emergency landings due to bad weather: especially fog, wind and rain.

At one stage, she took the decision to rise up through a bank of fog in the hope that she would soon reach the top and be able to see where she was going; at more than 750 metres (2,500 feet) she was still in a white-out, and when she finally did emerge, she had icicles forming on her eyelashes. 'That would have been the most terrifying moment of my life,' she recalled, interviewed on Swedish television, 'but all the birds were up their migrating with me; so I knew I was going in the right direction!'

Sacha's brave and audacious plan worked: she won the support of, among others, Sir David Attenborough and her distant relative Dame Judi Dench, appeared frequently on TV and radio, and featured in many newspapers and magazines. She also became a star on social media, as her faithful followers traced her three-month journey, until she flew above the White Cliffs of Dover to reach Britain in late November. The resulting publicity from her flight, both at home and abroad, kick-started co-operation between nations and conservation organisations all along the swans' route.

In mid-December, by the end of her epic flight, when Sacha finally landed at Slimbridge along with Maisie and the other birds, she had rightly earned her title of 'the human swan'.

Back on my home patch of the Somerset Levels, a more than usually wet autumn has led to widespread flooding, cutting off several minor roads and turning fields into temporary lakes.

As happens every winter, both the permanent and transitory waterbodies attract a wide range of wildfowl. Wigeon, resplendent in their smart plumage of pink, grey, chestnut and yellow-ochre, come here in their thousands from Iceland, Scandinavia and as far away as Siberia, where they breed alongside both whooper and Bewick's swans. Heavy-billed shoveler feed alongside diminutive teal, while pintail – surely our most elegant duck – glide along in their serene beauty. Mute swans are everywhere: loafing around in grassy fields, swimming along rhynes and drains, and from time to time flying overhead with whooshing wings.

MUTE SWAN.

Once, they would have been joined by flocks of wild swans: large, rangy whoopers, along with the smaller and more compact Bewick's, used to overwinter regularly in Somerset. There is an intriguing record of a 'whistling or wild swan' which was shot and wounded by 'a gentleman some miles from Bridgwater' as early as 1806. Said to be no larger than a goose, this was almost certainly a Bewick's swan, more than two decades before the species was officially discovered and named.

In February 1974 an extraordinary total of 386 Bewick's swans was recorded on King's Moor and West Sedge Moor, on the southern half of the levels; since the turn of the millennium, however, flocks have rarely reached double figures. Whoopers have never been so common here, but their numbers have also declined considerably in recent years.

Since we moved here to Somerset in 2006, both Bewick's and whooper swans have become less and less regular; some years now go by with just a handful of records of each species. So, I was pleased when, as I scanned the flooded Curry Moor one January morning, I saw a party of seven whooper swans – four adults and three young – feeding alongside the mutes. It was heart-warming to see these classic birds of winter, at a time when so many species are no longer coming south.

CYGNUS FERUS.

EPILOGUE

Towards the end of February, only a couple of days before the end of the meteorological winter, I took a trip to the best-known place to see swans in the world: Abbotsbury Swannery in Dorset. I had timed my visit well: after a week or more of blustery wind and rain, the day dawned bright and sunny, with an accompanying chorus of birdsong.

The swans also seemed to appreciate the change in the weather: many had already paired up and were beginning their courtship display. The sight of a male and female bending necks and joining bills to form the famous 'heart shape', backlit by the morning sun glinting off the water, told me spring really was just around the corner.

There were plenty of other birds here too: tufted ducks and pochards, various species of gull, several wintering whoopers and one resident black swan. But these were simply overwhelmed by the sheer number of mute swans – hundreds in all – spread out across the water or hauled up on the surrounding land. Seeing this fiercely territorial species being so accepting of the others was certainly a novel experience.

My guide was Steve Groves, the swanherd at Abbotsbury, who

has worked here for more than thirty years. Steve is the latest of a long line of swanherds, for the swannery has a distinguished history, going back almost a millennium, to before the Norman Conquest. Sometime during the 1040s, a group of Benedictine monks chose this site on the Fleet lagoon for their monastery, on land that had been granted by King Cnut (Canute) before his death a few years earlier.

The lagoon – the largest in Europe – is a shallow brackish pool, protected from the sea by the famous Chesil Beach, a vast bank of shingle stretching along the Dorset coast. Some years after the monastery was established, the monks began to keep a herd of swans to supply meat for their banquets.

Exactly when this happened is hard to say. We do know that William Squilor was appointed swanherd of Abbotsbury in 1393, towards the end of Richard II's reign; whether he was the earliest to hold that post, or simply the first to be recorded, is not known. We also know that in 1541 – almost 500 years after the holy house was first established – it was brutally destroyed, following Henry VIII's notorious dissolution of the monasteries.

For many monastic communities, the dissolution brought their traditions to an abrupt end. Fortunately, however, in 1543 the site and its surrounding lands were bought by Sir Giles Strangways, who continued the tradition of keeping swans here. His descendants (later the Earls of Ilchester) still own and manage the swannery today, fifteen generations down the family line.

Because of the family's long history as owners of the swans, they have the right to put a distinguishing ring on the leg of each cygnet that hatches here. Recoveries of Abbotsbury's swans have been made as far away as the Loire Valley in western France, as well as at various locations closer to home.

One reason Abbotsbury thrived for so long, where other medieval swanneries dwindled and disappeared, is that the site itself is absolutely ideal for mute swans. They have freshwater pools where they choose to nest, as the cygnets are intolerant of salt water; but for feeding, the optimum habitat is the brackish waters of the Fleet, which provide plenty of *Zostera*, or eelgrass, on which the birds thrive. Abbotsbury's swans are also given supplementary food: twice a day in winter, and three times a day when the swannery is open during the spring, summer and autumn – including daily public feeds which the visitors can join.

The swans here are not just given a helping hand with food: the reeds around the edge of the site are also cut regularly to provide nesting material. Steve explained that the birds do not carry their nesting material any distance, so bundles of reeds are strategically placed around the site to enable them to build their vast nests more easily.

When they do start breeding, most pairs are less territorial than they would be elsewhere; they need to be, given that they are often nesting cheek by jowl with another pair. The one exception to that rule occurs when two unmated males decide to pair up; because they don't have any nesting duties, they tend to be more aggressive than fully paired males, chasing all the surrounding birds away. As Steve says, 'They can be a bit of a pain!'

Breeding success does vary considerably from year to year: typically, between fifty and a hundred pairs attempt to nest, but many are unsuccessful. 'If we get two hundred cygnets fledged

in a season we are delighted,' Steve told me. 'If it's below a hundred, that's a bad year.' An electric fence surrounds the nesting area, to keep foxes and badgers out; but mink and rats can wreak havoc, and great black-backed gulls sometimes take eggs and young. Ironically, the biggest danger to the youngsters comes from the other adult swans, which will sometimes attack and kill a cygnet that strays into their territory.

As we were chatting, a disturbance flared up on the water in front of us. One male swan had taken exception to the proximity of another. Spreading his wings and beating them loudly on the surface, he flapped clumsily towards the intruder, to chase him away.

Nor are the adult swans here especially aggressive towards their offspring; unlike the Bishop's Palace birds in Wells, which start to chase away their cygnets the autumn after they hatch, Abbotsbury's swans are far more easy-going.

But they are not so tolerant of their keepers. Steve told me horror stories of colleagues who have been attacked by swans, including one who was knocked unconscious when a swan pushed him over with a beat of its wing. 'They use the joint at the wrist to inflict the blow, and this can produce some nasty bruises.' Maybe that urban myth that a swan 'can break a man's arm' might contain a grain of truth, after all.

Oddly, the swans completely ignore visitors, and hardly ever show any aggression towards them, but they can identify the staff by sight – 'They know who we are and they hate us with a passion!' Steve explained that this may be because the staff have to lift the nesting swans off the nest each morning to record the progress of egg-laying and incubation. 'So, even though we provide them with bed and breakfast, they still see us as a threat.'

Each summer, after breeding, the swans become flightless for a short period as they moult their wing feathers. Every two years, at this time, they are rounded up to check their weight and measurements and general health.

In winter, more swans arrive at the swannery, increasing the total numbers: in 1980 a record number of 1,238 birds was counted. But in recent years the wintering population has dropped significantly – almost certainly because of milder winters, along with an increase in alternative wetland habitats, meaning that food is now available elsewhere all year round.

As we were watching the swans, I noticed a sign illustrated with a photograph of a company of ballet dancers, in full costume, performing behind a flock of rather bemused-looking swans. The caption reads:

> Anna Pavlova studied the swans here while rehearsing her legendary performance of *Swan Lake* in the late 1920s. The photographer stood near here as members of the corps de ballet danced *Swan Lake* at the Swannery.

It's a lovely story, but as Steve pointed out, the photograph was clearly taken much later, in the 1950s, long after Pavlova's death, during a re-enactment of her visit. But it seems that she did visit the swannery; sadly, the momentous event was never captured on camera. Just one of the many stories about swans that turn out to be a mixture of truth and falsehood.

While writing this book, I became aware how many people on social media post photographs and tell stories about 'their' swans: celebrating the laying of the eggs, the birth of the cygnets and the majesty of the adults, while rightly condemning those who – through deliberate actions or simple carelessness – threaten their lives. They also use swans as a way of cheering up themselves and their followers, as in this plaintive posting, accompanied by a photo, in March 2021:

> Crappy unproductive day of gloom; going to write it off and do some knitting. Try again tomorrow.
> For now, here's a swan building a nest.

No wonder Abbotsbury Swannery attracts more than 50,000 visitors every year, from the UK and abroad. Nowhere else in the world can you see so many mute swans squeezed into such a small space. It is a truly unique spectacle.

Acknowledgements

Many people have helped me with writing this book, and in doing so have shown both their passion for swans and their generosity in sharing their immense knowledge and experience with me.

Regarding the mute swan, I particularly wish to thank Professor Christopher Perrins; my fellow author Helen Macdonald, for kind permission to quote from her essay on swan-upping in *Vesper Flights*; another fellow nature writer, Donald S. Murray; Moira Anderson and Rosemary Cooke from the Bishop's Palace, Wells; Steve Groves and Charlie Wheeler from Abbotsbury Swannery; Melanie Nelson from the Swan Sanctuary; David and Venetia Hopkins; Richard and Dylan Tripp; my colleague at Bath Spa Dr Terry Gifford; Professor Axel Goodbody; wildlife cameraman Mark Payne-Gill; and two of my students on the Bath Spa MA Travel and Nature Writing, Gayle Wood and Ruth Lawrence – Ruth's lovely poem serves as the epigraph to this book.

Regarding whooper and Bewick's swans, I would like to thank Mark Simpson and Scott Petrek of the Wildfowl and Wetlands Trust at Slimbridge; and Dr Eileen Rees, Sacha Dench and Dafila Scott for sharing their expertise and stories about these very special birds. Eileen Rees also kindly read through the Winter chapter and made many helpful suggestions.

As always, the team at Square Peg (Penguin Random House) have done a great job on the book: Mireille Harper and Maxine Sibihwana in Editorial; Ryan Bowes in Publicity; Jane Kirby in the rights team; and the designers and production team, Rosie Palmer, Lily Richards and Shabana Cho, who also found the lovely cover image by Simon Turvey. Thanks, as ever, to my editor Graham Coster and my agent Broo Doherty. And finally, to my wife Suzanne, who shared the joy of discovering our local mute swans' nest during that memorable lockdown spring of 2020.

List of Illustrations

(1921–1930) 2nd edn. *(Biodiversity Heritage Library, courtesy of Smithsonian Libraries)*

p. 31 *Ornithologia neerlandica. De vogels van Nederland* by Koekkoek, M. A. (Marianus Adrianus), illustrator van Oort, E.D. (1922) *(© Biodiversity Heritage Library, courtesy of Smithsonian Libraries)*

p. 38 *Domestic Swan* (c. 1832) *(©Interfoto / Mary Evans Picture Library)*

p. 44 *Cygnus atratus, Black Swan (The Birds of Australia)* by Gould, E., Gould, J. and Richter, H.C. (1821–1902) *(© Biodiversity Heritage Library, courtesy of Smithsonian Libraries)*

p. 49 *Cygnus atratus, black swan, Watercolour 351 by the Port Jackson Painter from the Watling Collection* (c.1700) *(© The Natural History Museum / Alamy Stock Photo)*

p. 53 *Anna Pavlova in the role of the Dying Swan* (c.1905) (black and white photo) *(© Private Collection / Bridgeman Images)*

p. 55 *Black and white swan* (c.1910) *(© Quagga Media / Alamy Stock Photo)*

p. 60 *Swans Nest* (gouache on paper) (c.1900) *(© Look and Learn / Bridgeman Images)*

p. 64 *Two swans,* by Wenceslaus (c.1600) (etching) *(© Iconographic Archive / Alamy Stock Photo)*

p. 70 *Eggs from* The Book of Household Management by Beeton, I.M. (1892) (engraving) *(© British Library / Bridgeman Images)*

p. 72 *Two swans chasing a fox from their nest on a shoreline* by Tischbein, J.H.W. (c.1810) (pen and brown ink with coloured wash over black chalk) *(© Private Collection / Bridgeman Images)*

p. 80 *Swan being chased by three dogs* by Hondius, A.D. (1668) *(© Private Collection / Bridgeman Images)*

p. 155 *Swan Vestas Packaging* (c.1930) *(© The Advertising Archives / Bridgeman Images)*

p. 160 *Frozen Out* by Leslie, G.D. (c.1866) (oil on canvas) *(© Christie's Images / Bridgeman Images)*

p. 164 *Cygnus, Cygnus* from *Naturgeschichte der Vögel Mitteleuropa* by Hennicke, C.R., Naumannn, J.A. and Naumannn, J.F. (1901) *(© Biodiversity Heritage Library, courtesy of Smithsonian Libraries)*

p. 172 *Cygnus Bewickii Yarrell* from *Naturgeschichte der Vögel Mitteleuropa* by Hennicke, C.R., Naumannn, J.A. and Naumannn, J.F. (1901) *(© Biodiversity Heritage Library, courtesy of Smithsonian Libraries)*

p. 181 *Bewick's Swan* from *Coloured figures of the birds of the British Islands* by Lilford, T.L.P., Keulemans, J.G., Salvin, O., Newton, A. and Thorburn, A. (1885–1897) *(© Biodiversity Heritage Library, courtesy of Smithsonian Libraries)*

p. 187 *Swans* from *Illustrations from British Game Birds and Wildfowl* by Morris, B.R. (1891) (colour lithograph) *(© Look and Learn / Bridgeman Images)*

p. 188 *Mute Swan (Cygnus olor)* from *A History of British Birds* by Morris, F.O. *(© Mary Evans Picture Library)*

p. 190 *Cygnus Cygnus* from *The Birds of Great Britain* by Gould, J., Vol 5. (1873) *(© Natural History Museum, London / Bridgeman Images)*

p. 196 *The Rivals* by Tunnicliffe, C.F. (watercolour on paper) *(© Newport Museum and Art Gallery / Bridgeman Images)*

p. 200 *Swan on Lake* by Khouja, F. (2004) (pastel) *(© Faisal Khouja / Bridgeman Images)*